Healable Polymer Systems

Fundamentals, Synthesis and Applications

2nd Edition

Polymer Chemistry Series

Editor-in-chief:
Alaa Abd-El-Aziz, *Harbin Engineering University, China*

Series editors:
Céline Calvino, *University of Freiburg, Germany*
Jianhua Dong, *Beijing Forest University, China*
Tanja Junkers, *Monash University, Australia*
Wai-Yeung Wong, *The Hong Kong Polytechnic University, Hong Kong*

For a list of titles in this series see: rsc.li/polymer-chemistry

How to obtain future titles on publication:
A standing order plan is available for this series. A standing order will bring delivery of each new volume immediately on publication.

For further information please contact:
Book Sales Department, Royal Society of Chemistry, Thomas Graham House, Science Park, Milton Road, Cambridge, CB4 0WF, UK
Telephone: +44 (0)1223 420066, Fax: +44 (0)1223 420247
Email: booksales@rsc.org
Visit our website at books.rsc.org

Healable Polymer Systems
Fundamentals, Synthesis and Applications
2nd Edition

Edited by

Barnaby W. Greenland
University of Sussex, UK
Email: b.w.greenland@sussex.ac.uk

and

Wayne Hayes
University of Reading, UK
Email: w.c.hayes@reading.ac.uk

ROYAL SOCIETY OF CHEMISTRY

Polymer Chemistry Series No. 41

Hardback ISBN: 978-1-83767-044-4
PDF ISBN: 978-1-83767-353-7
EPUB ISBN: 978-1-83767-354-4
Print ISSN: 2044-0790
Electronic ISSN: 2044-0804

A catalogue record for this book is available from the British Library

The Royal Society of Chemistry is a charity, registered in England and Wales, Number 207890, and a company incorporated in England by Royal Charter (Registered No. RC000524), registered office: Burlington House, Piccadilly, London W1J 0BA, UK, Telephone: +44 (0)20 7437 8656.

For further information see our website at www.rsc.org

For general enquiries, please contact books@rsc.org

For EU product safety enquiries, please email books@rsc.org or contact Royal Society of Chemistry Worldwide (Germany) GmbH, Römischer Hof, Unter den Linden 10, 10117 Berlin.

Printed in the United Kingdom by CPI Group (UK) Ltd, Croydon, CR0 4YY, UK

Preface

It is without doubt that the synthesis and large-scale production of polymeric materials has had a tremendous impact upon society in the last century. During this period, polymer chemists have refined their art and learnt how to manipulate reactions and tune processing strategies to make new materials with precise and useful properties. Reliable polymer synthesis has enabled the proliferation of many everyday items that are taken for granted, ranging from polymeric water pipes to carry clean water to our homes to the complex block co-polymers that are required in the production of silicon microchips.

However, as we move deeper into the 21st century, and especially in the decade since the publication of the first edition of this monograph, mankind finds itself in a new environment. The era of cheap petroleum which has enabled the provision of inexpensive monomeric feedstocks and energy is rapidly drawing to a close. Thus, there is an ever greater need to examine routes to polymeric materials that can be recycled, re-healed and mended in order re-attain the desired function of an object without needing to replace either single components or the entire device.

In common with the first edition, the primary objective of this monograph is to provide the reader with a manageable reference text on healable polymer systems that have been reported to date, principally from the perspective of the practicing synthetic organic chemist. However, in order to understand the area effectively, it is important that any scientist seeking to enter this field has a general grounding in the evaluation of healable polymers. Therefore, this monograph opens with an overview of the key parameters and terminology used to describe healing polymeric materials together with a brief outline of the major assessment methods that are used to demonstrate healing behavior. Subsequently, the book is split into five chapters, written by leading specialists in their fields, that cover the

---

Polymer Chemistry Series No. 41
Healable Polymer Systems: Fundamentals, Synthesis and Applications, 2nd Edition
Edited by Barnaby W. Greenland and Wayne Hayes

Published by the Royal Society of Chemistry, www.rsc.org

development of each area of healable polymer synthesis up to the present date. A new feature of this edition is the addition of specific examples of the applications of these healable materials in 'real-world' settings. The aim of this is to exemplify to the reader how these fascinating materials are moving from laboratory curiosities towards commercially available products that have a tangible impact on society.

Encapsulated monomer approach to healable polymers and composites: this chapter provides an overview of how the encapsulation, release and polymerization of monomers within polymer composites has been used to create healable materials. The review also summarizes key mechanistic studies using exemplars from the literature.

Reversible covalent bond formation as a strategy for healable polymers: highly efficient and reversible covalent bond forming reactions such as the Diels–Alder cycloaddition or thiol-ene reactions have been harnessed to produce healable polymeric materials and chemically adaptive networks. This chapter summarizes the most important examples of healable polymeric materials that make use of reversible covalent bond forming reactions to recover their strength after damage.

Healable supramolecular polymeric materials: this chapter reviews the use of weak non-covalent interactions such as electrostatic forces, hydrogen bonding, metal–ligand interactions and π–π stacking to assemble healable polymer networks and nano-composite materials.

Thermodynamics of self-healing reactions and their applications in polymeric materials: this chapter considers the thermodynamic parameters needed to produce successful healable polymeric materials. Entropic and enthalpic contributions to the Gibbs free energy during healing and self-healing are considered in the context of the mechanisms within the polymeric material that contribute to effective healing.

We are extremely grateful to the authors of each chapter, and it has been a pleasure to work with each in the preparation of this edition. Finally, we would like to thank our families (Emma, William and Benji & Kerry and Louisa) for their patience and understanding during the assembly and editing of this monograph.

Barnaby W. Greenland and Wayne Hayes

Contents

Polymer Chemistry Series No. 41
Healable Polymer Systems: Fundamentals, Synthesis and Applications, 2nd Edition
Edited by Barnaby W. Greenland and Wayne Hayes

Published by the Royal Society of Chemistry, www.rsc.org

CHAPTER 1

Healable Polymeric Materials

BARNABY W. GREENLAND*[a] AND WAYNE HAYES[b]

[a] Department of Chemistry, University of Sussex, Falmer Campus, Sussex, BN1 9RH, UK; [b] Department of Chemistry, The University of Reading, Whiteknights, Reading, RG6 6AD, UK
*Email: b.w.greenland@sussex.ac.uk

1.1 Introduction

Throughout time, technological progress has served to improve living standards and increase the life expectancy of mankind. This progress is linked directly to new materials that humans have learnt to master and manipulate. The connection between the basic fabrics that tools can be made from and the progression in human development has become so intertwined that these advances have come to define specific eras: the stone age, bronze age and iron age. For the past 50 years or so, the role that silicon has played in advancing people's ability to address significant challenges *via* computer technologies cannot be underestimated (ranging from the development of personal computers to the evolution of artificial intelligence). There is also a growing consensus that the current period may come to be known at the *Plastic Age.*[1] From the seminal discoveries of Staudinger[2] and Caruthers[3] in polymer science during the 1930s, carbon and inorganic-based macromolecular/polymeric products have proliferated through the modern world finding applications in all areas from inexpensive disposable packing materials to life-enhancing hip replacements and life-saving body armor.

In 2021 sales of raw polymeric materials exceeded €121.5Bn in the European Union.[4] With demand for polymeric products growing even whilst the cost of crude oil rises there is a clear need to move away from the culture of disposable products that society has become accustomed to, and to deploy

Polymer Chemistry Series No. 41
Healable Polymer Systems: Fundamentals, Synthesis and Applications, 2nd Edition
Edited by Barnaby W. Greenland and Wayne Hayes

Published by the Royal Society of Chemistry, www.rsc.org

materials that are sustainable.[5] This can be achieved one of several ways: producing more polymers from renewable feedstocks; increasing the proportion of recycled polymers in circulation[6–8] and by increasing the lifespan of the polymeric products. It is these latter two potential solutions in which healable materials have the most to offer.[9–18] As shall be discussed in later chapters (see Chapters 3 and 4), the reversible nature of both covalent[19,20] and supramolecular bonds[21,22] is frequently exploited in producing healable materials that lend themselves to efficiently recyclable materials. More clearly, producing materials that can heal either small cracks or major fractures will have a significant impact on the longevity of a host of polymeric products from everyday devices such as mobile phones,[23] bulk materials as used in the construction industry, through to structural components of airplanes. Indeed, the potential impact of healable polymers on modern society is significant with the market value of such materials predicted to grow rapidly to ~$70bn by 2032.[24]

The diverse nature of applications for polymeric materials has occurred because of the chemist's increasing ability to design and synthesize new monomers and polymeric architectures (for example: multi block co-polymers, branched or network materials), delivering products with useful functionalities and physical properties. As synthetic techniques have progressed, the materials scientist's 'toolset' for characterizing the materials at the micrometer, nanometer and angstrom scales has also improved – so the ability to successfully predict and measure the properties of new polymeric materials improves year-on-year. This structure/property interplay serves to increase the speed that innovative materials with step-changing properties can be conceived, produced and brought to the market – further enhancing the modern world. In addition, new manufacturing techniques have been developed in the past decade such as additive manufacturing (commonly referred to as 3D printing) that permits complex 3D structures with complex polymeric compositions to be generated with precision in order to fulfil a specific function.[25]

In the formative years of the field of polymer science in the late 1930s and early 1940s, the primary driving force of the research carried out was to produce new materials whose properties (*i.e.* strength and thermal stability) were suitable for producing inexpensive items that could be mass produced – *i.e.* polystyrene cups, PET water bottles, strong fibers for rope, vulcanized rubber for car tires and many of these polymers were utilized in World War Two. Recently, however, research has focused on producing high value items from polymers which necessitates new macromolecules with novel properties, for example, shape memory materials, photo-conducting or luminescence devices and composite polymers that exhibit strength surpassing that of the strongest metal alloys.

By their very nature, studies that afford such polymeric materials frequently require a multidisciplinary approach. This can require close collaboration between synthetic chemists who can generate and manipulate the materials that differ at the atomistic level and engineers who build the device to test the product (*i.e.* a working solar cell[26] or printed circuit board[27]).

Optimization of these complex systems necessitates multiple iterations taking many person-years of effort before finally settling upon a suitable balance between the time, raw material, synthetic and fabrication costs required to attain the desired design criteria. Modern high-throughput screening approaches,[28,29] machine learning[30–33] as well as 3D printing prototyping are now being employed to diminish the time taken to process lead candidates through the discovery→development→production process to progress organic-based products to the marketplace.

This inter-disciplinary work ethos is clearly applicable to the field of healable polymer research. The most basic definition of a healable material requires a polymer with a given strength to be damaged, reducing its physical properties, and then at a later point to have regained some of its lost strength. Any research plan will require at least:

i. The design, synthesis and characterization of new polymers (note: large scale syntheses are no longer required – many high-throughput screening assays can test small quantities of material (*ca.* mg)) to ascertain key physical parameters;
ii. A method to fabricate a sample suitable for mechanical testing;
iii. A mechanical test to assess the pristine sample and healing nature of the material and
iv. An iterative development cycle whereby data from the mechanical testing will feedback into the production of the next generation of materials. Recent advances in automated synthesis, analysis and data processing have led to dramatic improvements in the rate that new materials with desirable properties can be generated.[28,29]

In order to complete the iterative development loop successfully, from a chemist's perspective, it is important that whilst the chemistry of the new material is fully understood it is also imperative to have an understanding of the mechanical tests required to demonstrate healing and the terminology and conventions used by engineers. After summarizing the potential benefits of producing commercially viable healable polymers this chapter will define the requirements and different categories of healable materials before providing an overview, aimed at the practicing chemist, concerning the techniques currently employed to study the healability of these fascinating new materials.

1.2 Healable Polymers – Potential Applications

While producing healable materials is an intuitively rewarding challenge, it is worth considering why developing healable materials may be beneficial in differing circumstances. Polymeric materials are now used across a wide range of applications and fulfil a multitude of roles. In many cases the polymer itself may play little part in adding to the strength of the product but may simply provide an attractive finish – for example, the outer surface coating on a modern car. In these circumstances, small scratches caused by wear and tear are unlikely to alter the safety of the passengers. However, coatings that can regain their original luster after damage are of considerable commercial importance from an aesthetic viewpoint and improved corrosion resistance. Indeed, healable coatings have been put into production in recent years by manufacturers such as Nippon Paint[34] or Bayer[35] and the market for materials of this type is projected to grow to over \$10bn by 2028.[36]

Focusing on higher strength materials, one can envisage the benefit of healable gas and water pipes. In these instances, producing new materials that can autonomously fix fractures, *in situ*, without the need to dig up the roads would be a clear advantage to consumers and suppliers alike. There are also more high profile (although currently lower volume) applications associated with the high strength polymers and composites that are used in performance cars, military aircraft and the current generation of passenger airplanes. For example, significant aspects of the airframes and other key structural elements of modern commercial aircraft (such as the Airbus A380 'Superjumbo' and Boeing's 787 'Dreamliner') are polymer composite in nature. In these situations where safety is an overriding design criterion, it might be expected that producing a polymer that can completely regain its strength after damage would be highly desirable. The ability of a material to self-heal in an autonomous manner is especially important in composite materials where detecting damage within the interior of these structures is inherently difficult (and less well-developed) than with identification of damage sites in conventional aircraft made from metal-alloy airframes.

In all these situations a number of restraining factors must be considered, as outlined in the following sections.

1.2.1 Cost

In the majority of cases, new technologies decrease in cost as they become more established and are produced in greater scale. From the outset, healable materials could also offset higher amounts in terms of initial outlay by increasing longevity and lowering (or even serving to eliminate) routine maintenance costs, which are especially high in safety-critical situations such as transport.

1.2.2 Performance When Compared to Conventional Materials

Introducing healability into a functional component must not compromise the strength or increase the density of the component over the original. A healing polymer that is double the density of its conventional (non-healing) predecessor will not be suitable for many applications such as transport, where the energy required to move the heavy healing system will be seen to outweigh any potential safety advantage. This is particularly important when considering the mode of healing at the molecular level. Conceptually, introducing latent, unpolymerized monomers and functionalities into a material that can form new bonds, to heal a fracture is a pleasing potential solution – but one must consider how much stronger the material would have been if the latent functionalities were fully polymerized in the original formulation.[7]

1.2.3 Sustainability of Materials

Whether the material can be produced from sustainable feedstocks and subsequently recycled or repurposed at the end of its useful life with minimal energy input, should be considered.

1.3 Categorization of Healable Materials

As noted above – the simplest criterion of a healable material is that it is able to regain its strength after a damage event. This can be visualized by plotting the strength of the polymer as a function of time (shown as Tensile modulus in Figure 1.1).

In this simplified example, the sample remains undamaged between time point A and B indicated by the constant strength of the system. At time point B the material is damaged, resulting in an instantaneous loss of strength (point C on the plot). Subsequently, the sample heals, and the strength of the material increases to match that of the pristine, undamaged sample (time point D). The rate of healing is indicated by the gradient of the slope between time points C and D. After the healing event, the strength remains constant for the remainder of the time.

A truly ideal healable material may have additional properties, for example, it may be able to heal on multiple occasions (Figure 1.2). These break/heal cycles may have several effects on the material. Figure 1.2(A) shows the strength *versus* time plot for an ideal sample that can regain its strength completely over three break/heal cycles. Figure 1.2(B) shows the situation encountered more frequently, whereby the sample only partially recovers its strength after each damage event. Thus, ultimately, the

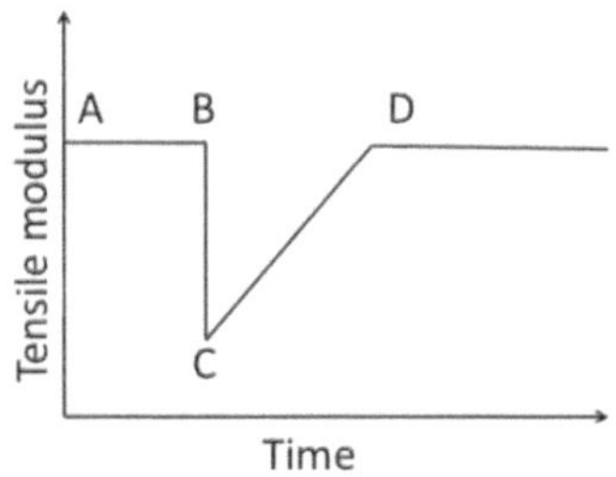

Figure 1.1 A schematic plot of tensile modulus as a function of time for a healable material undergoing a single break and heal cycle.

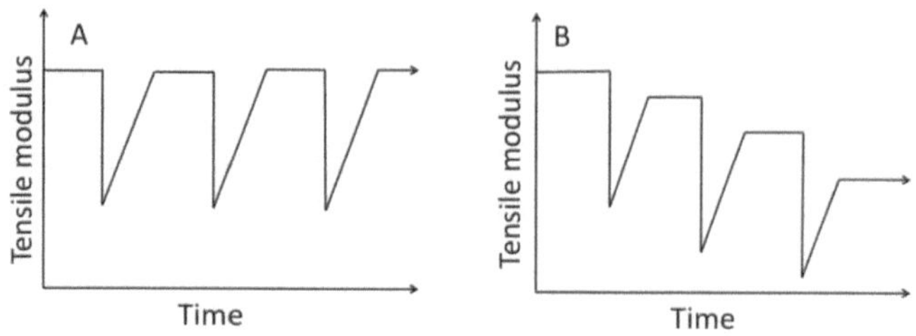

Figure 1.2 Schematic plots of tensile modulus as a function of time for a healable material undergoing multiple break/heal cycles: Plot (A) a material that completely recovers the strength of the pristine material. Plot (B) a material capable of only recovering a fraction of the strength lost during the damage event.

performance of the sample will degrade, but depending on the timescales involved, this may be adequate to fulfil the typical lifecycle of the product.

This simple treatment immediately raises several pertinent issues concerning healable material characterization, some of which are highlighted below:

i. How strong is the pristine material?
ii. How significantly can the material withstand damage yet still perform adequately with respect to its use?
iii. How quickly can it regain its strength?
iv. Under what conditions does the material heal?
v. How many break/heal cycles can the material undergo before the performance falls away?
vi. How easily can the site of damage in the polymer-based assembly be accessed?

The target answer to many of these questions will be application dependent. For example, the strength and number of healing cycles that a mobile phone screen or case may be expected to pass through in a typical lifetime will be very different from that of a polymer component embedded deep within the wing of an airplane. However, a key universal consideration is what parameters should be measured to quantify healing in a polymeric system, and how should they be reported. The following two sections provide an overview of these two areas and are presented from the synthetic chemistry perspective to provide insight into some of the most common techniques and measurements reported for this rapidly evolving field of research.

1.4 Healing – Definitions

Although the initial design criteria for a healable material is the production of a component that can completely regain its physical properties after sustaining damage, there are many practical situations where this may not be possible or indeed necessary. In these situations, it is necessary to quantify the loss in performance in the healed material. This is most frequently expressed as the 'healing efficiency' (η_{eff}) of the material. η_{eff} is a dimensionless parameter which is the ratio of a specific mechanical property of the material measured before and after healing, and is expressed as a percentage (eqn (1.1)).

$$\text{Healing efficiency, } \eta_{\text{eff}} = \left[\frac{\text{Mechanical value (healed)}}{\text{Mechanical value (pristine)}}\right] \times 100 \qquad (1.1)$$

Thus, materials with η_{eff} approaching 100% (note: it is possible to have $\eta_{\text{eff}} > 100\%$) have regained the strength of the pristine material, whereas low values for η_{eff} indicate that the material has not regained its original strength significantly. Healing efficiency is, therefore, an easy parameter to obtain and clearly indicates the success of a healing process, but the simplicity of this term can lead to misleading results, especially when only a single healing parameter is measured. In a typical break/heal experiment, a sample will be stretched to its breaking point and a variety of mechanical properties are measured during such an experiment, for example: tensile modulus, elongation to break and modulus of toughness. After healing, the sample may exhibit (and indeed frequently does exhibit) a higher η_{eff} for one of these parameters when compared to the other two. Thus, only by measuring and quoting η_{eff} for all of the appropriate parameters for a specific polymer can a true assessment of the nature of the healable material be made.

A further complication for a researcher new to this field is that there is not a standard definition of what actually constitutes a *damaged* material. In this respect, papers describe η_{eff} values that may have been calculated for the healing of a single microscopic crack in a bulk polymer, or alternatively after a material has been physically broken into multiple sections and the parts then separated and rejoined prior to obtaining the healing efficiency data. To this end, the term 'healing' serves to cover several repair scenarios and is evidently subject specific – readers of articles in this area must always bear this caveat in mind.

1.5 Measuring Healing

Testing procedures for healable polymers have been inspired by the well-established area of materials chemistry and engineering and broadly fall into two categories:

i. Mechanical load testing where a monotonic force is applied.
ii. Rheological tests where an oscillating force is applied.

Both these experimental protocols have been used to assess the healing properties of new polymeric materials. The most frequently studied of these techniques are covered in the following sections and also in Chapter 2.

1.5.1 Cantilever Beam Tests to Determine Healing

Cantilever beam (CB) tests are suited to testing non-elastomeric (brittle) samples, generally where the glass transition temperature (T_g) of the material is significantly above the testing temperature, or the polymer is highly cross-linked.[37]

Mechanical load testing requires fabrication of test samples which typically have maximum dimensions of the order 100 × 75 × 7 mm (typically requiring 50 grams of material). The geometry of the sample depends on the precise experiment (see Figure 1.3), with the four most common forms being:

1. Tapered double cantilever beam (TDCB);
2. Compact tension (CT);
3. Single edge notch bend (SENB) and
4. Single edge notch tension (SENT).

In each experiment the sample is prepared by manually forming a precrack, typically by scoring with a razor blade. The distance between the tip of the precrack and the applied force is termed $a_{pristine}$ (Figure 1.4). The sample

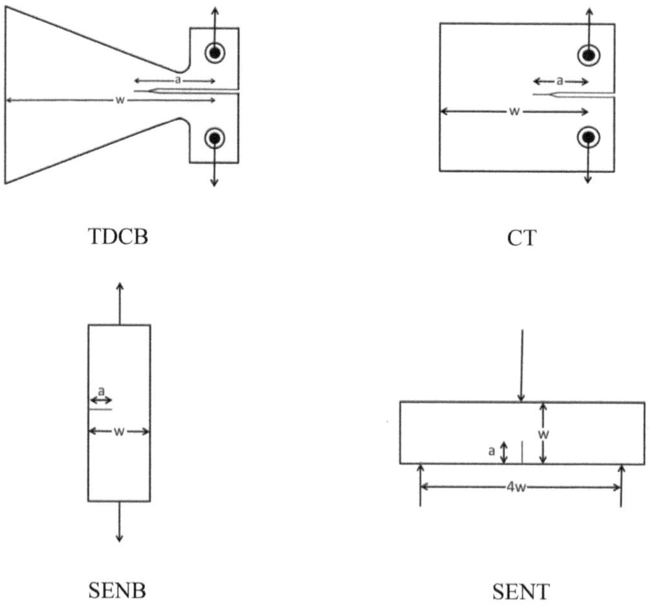

Figure 1.3 The four most common geometries for carrying out material property analysis on polymers. Key dimensions are indicated with double headed arrows: '*a*' is the precrack length, '*w*' is the most important geometric dimension. The position and direction of forces are shown with single headed arrows. Note: in the SENT experiment the total force on the top and bottom face of the material must be equal.

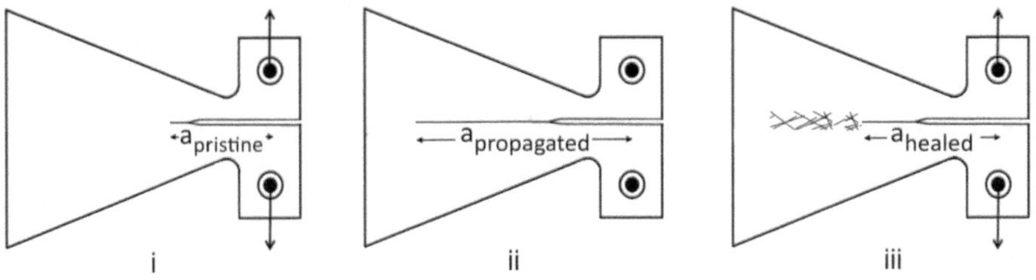

Figure 1.4 Typical break/heal experiment carried out on a sample with TDCB geometry. (i) Pristine sample with force being applied. (ii) Broken sample, crack has been extended, force removed to allow healing. (iii) The sample has healed, the crack reduced in length and the force is reapplied to measure the strength of the healed material.

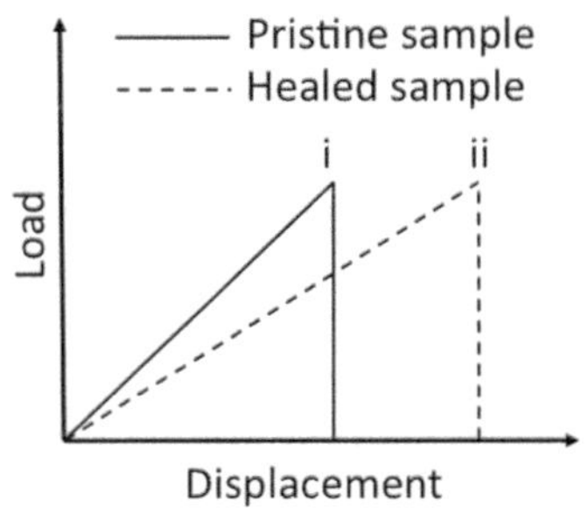

Figure 1.5 Typical load *versus* displacement plots for a brittle healable plastic in the TDCB geometry. The critical load for the pristine and healed samples (load at which they have broken) are labeled (i) and (ii), respectively.

is then subject to a force which induces a crack of length $a_{propagated}$. During this fracture event the mechanical properties of the pristine material can be obtained. The load is removed and the sample is then left to heal. Whilst unloaded, the fracture faces will come into close proximity, facilitating healing, over which period the crack reduces in length, resulting in a new distance between the fracture tip and the applied force (a_{healed}). The sample is then tested again and the healing efficiency calculated by comparison of the response of the pristine and healed samples.

The results from the healing experiment are presented as a plot demonstrating how applied load varies as a function of displacement, which is generally only a few millimeters. Representative load *versus* displacement plots for a break/heal experiment using samples in the TDCB geometry are shown in Figure 1.5.

It is evident from the plot in Figure 1.5 that the applied load is directly proportional to displacement up to a critical value for both the pristine and healed samples – beyond these values the samples break (as indicated at points (i) and (ii), Figure 1.5). The energy required to break the sample (defined as the *modulus of toughness*) is calculated by integration of the area under this displacement–load curve. In this example, where $a_{pristine} < a_{healed}$ (for example, see Figure 1.5) the healed sample exhibits a greater displacement before breaking. This results in an apparent increase in the modulus of toughness for the healed sample and a corresponding η_{eff} of greater than 100%. However, this result is a consequence of the increase in the length of the precrack between the two tests, which results in an increase in compliance in the sample.

TDCB geometry is a well-used method for investigating inelastic materials capable of exhibiting healable characteristics. This is primarily because the critical load (points i or ii in Figure 1.5, a typical healing parameter) for the pristine and healed systems is independent of the difference in length between $a_{pristine}$ and a_{healed} (as shown in Figure 1.5). In each of the other three testing geometries (CT, SEND and SENT) $a_{pristine}$ and a_{healed} must be measured to a high precision and any discrepancies accounted for. If these time consuming and difficult measurements are not made and $a_{pristine}$ is approximated to be equal to a_{healed} then large errors (*i.e.* >100%) are introduced into the results.

Load testing may also be carried out on elastic materials. In these cases the sample is formed into a typical 'dogbone' like shape, requiring from 50 mg up to multiple hundreds of grams of material depending upon the instrument used (see Figure 1.6). The increase in size of the sample at the termini serves ensures that the sample breaks cleanly in the middle, preventing the sample tearing where it is clamped to the instrument. Furthermore, the sample fracture at the clamped edges does not afford a representative indication of the strength of the sample and thus sample preparation is key to this assessment format. The dogbone is elongated and the response of the material measured and plotted as stress (load) *versus* displacement (Figure 1.6). Extension values for typical elastomeric materials

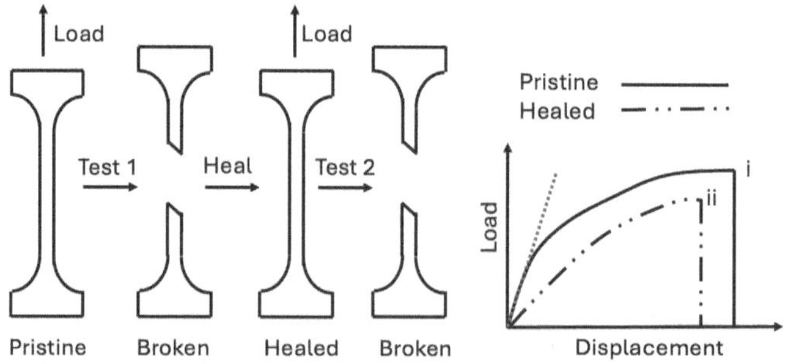

Figure 1.6 (Left) Schematic of a break heal test on a dogbone-shaped elastomer. The pristine sample is stretched to breaking point and the response measured (Test 1). The broken sample is healed and re-tested (Test 2). (Right) Typical displacement *versus* load plot for a pristine elastomer and healed sample. The ultimate tensile strengths of the two samples are labeled 'i' and 'ii', respectively. The Young's modulus of the sample is measured by taking the tangent to the initial slope (dotted red line for the pristine sample). The Modulus of Toughness is the area under the curves.

can be of the order of several 100% of the length of the pristine sample. The stress *versus* displacement plots reveal three key pieces of information about the sample which have been used as a measurement of healing efficiency:–

1. The *ultimate tensile strength (UTS)*, where the material breaks and the load response drops to zero.
2. The *stiffness* or *Young's Modulus* of the sample is measured by taking the tangent to the initial slope (see the dotted red line for the pristine sample in Figure 1.6).
3. The energy required to break the sample which is the area under each curve (referred to as the *modulus of toughness*).

1.5.2 Impact Testing

Impact testing is a well-studied technique for assessing the strength of both metals and polymers. During the test, the sample is completely severed by a falling or swinging weight from which the change in momentum of the system can be used to calculate the strength of the sample. Two of the most widely used variants of these experiments are the Izod and Charpy impact tests.[38,39] Sample sizes and geometries are of the order of $10 \times 10 \times 55$ mm and are defined by specific standards for ease of comparison (for example: North American Standards ASTM D256 or A370).

In both cases a notched, clamped sample is struck and broken by the head of a swinging pendulum, and the loss of energy of the system calculated by the difference in the height (angle) between the position that the pendulum began its motion and the angle it stops after breaking the sample. The two tests differ in that in the Izod test, the sample is clamped at one end of the long axis of the cuboid sample and the notch is on the same side as the impact. In contrast, samples for Charpy tests are clamped on both sides of the notch, which is on the opposite face to the impact (Figure 1.7).

The break/heal tests detailed in this section are conceptually easy to place into a real-world setting. In each test, the sample of bulk polymer is fractured or severed by applying a force, the sample is then left to heal and the material re-tested. However, many polymers, including linear polymers and supramolecular systems, are *visco-elastic* materials and as such are amenable to analysis by rheological techniques.

1.5.3 Rheological Healing Methods

Rheology is the study of the deformation and flow of matter, and is underpinned by an extensive mathematical framework, detailed in numerous excellent texts that are beyond the scope of this chapter.[40] The following sections will cover the important definitions, concepts and experiments that have been used widely to characterize and assess healable polymer systems.

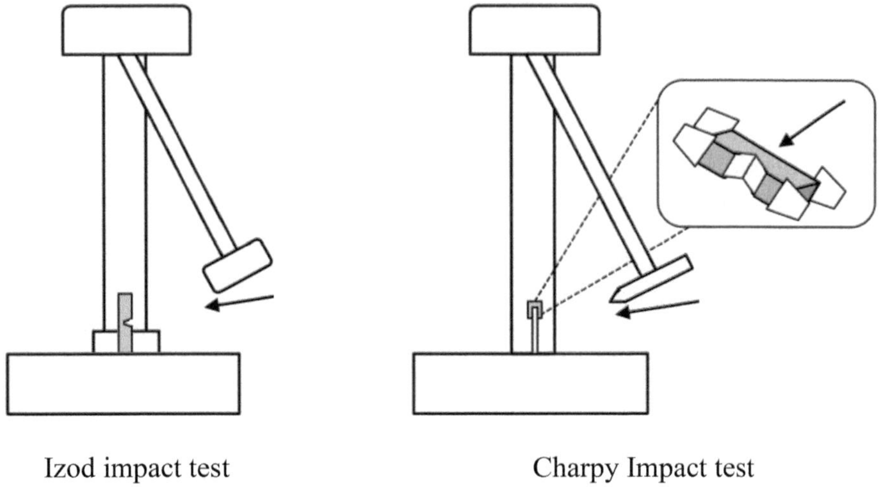

Figure 1.7 Schematic diagrams of the experimental set-ups for Izod and Charpy impact tests.

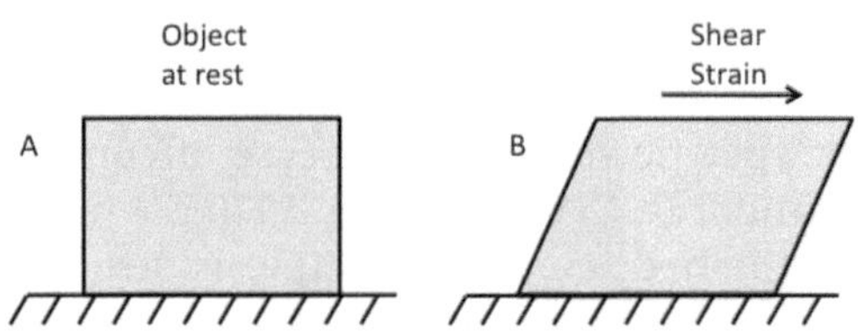

Figure 1.8 Schematic showing the strain response (displacement) of an object subjected to a shear strain.

1.5.4 Rheology – Background

All polymeric products are subject to forces of varying degrees throughout their lifecycle. It is therefore important to be able to model the strain (*i.e.* the displacement or change in shape) of these materials under a particular stress (*i.e.* the force applied to the polymer) – see Figure 1.8.

From the dawn of the modern polymer industry, it was apparent that polymers in the bulk, melt and solution states do not obey typical relationships between force and deformation. For example, many solids are ideally elastic in nature and obey Hooke's law (see eqn (1.2)) at small strains (deformations), in that once the strain is removed the materials recover their initial shape. Conversely, many fluids (encompassing liquids and gasses) follow Newtonian laws (see eqn (1.3)), whereby as the rate of strain increases, the stress (or drag) exerted by the fluid increases.

$$\tau = G\gamma \tag{1.2}$$

$$\tau = \eta(\mathrm{d}\gamma/\mathrm{d}t) \tag{1.3}$$

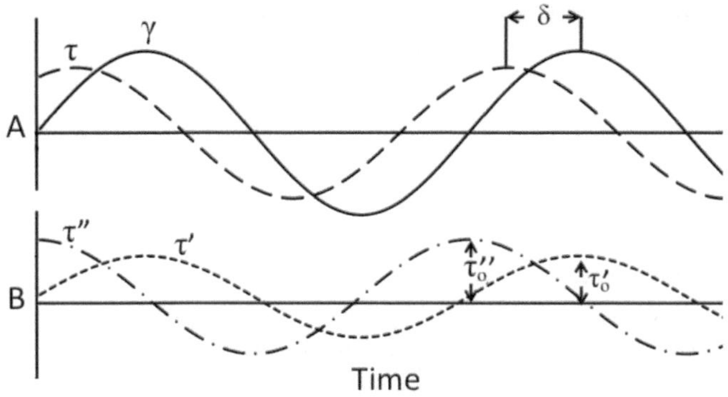

Figure 1.9 (A) Sinusoidally oscillating shear strain (solid line) produces a sinusoidally oscillating stress response time shifted by an amount termed δ. (B) Plots of the decomposed stress waves τ' and τ''. Note these waves have the same frequency as γ, with one in phase (τ') and one out of phase (τ'') with the input strain.

where: τ = stress, force per unit area. G = elastic modulus, a constant of proportionality. γ = strain, change in length; thus ($\mathrm{d}\gamma/\mathrm{d}t$) is rate of change in strain. η = viscosity, a constant of proportionality.

These two laws are inapplicable to polymers, which frequently exhibit a response to stress intermediate between those predicted by these two equations, and as such they are termed *viscoelastic* materials. The properties of viscoelastic materials are investigated *via* rheological experiments. During these tests an oscillatory force with a frequency of ω is applied to one face of the material (solid line, Figure 1.9(A)) and the response to this force is measured at the second, fixed face of the material (dashed line, Figure 1.9(A)). The stress response lags behind the strain by a measurable time period, termed δ. Thus, the stress and strain can be expressed as follows (eqn (1.4) and (1.5), respectively)

$$\gamma = \gamma_0 \sin \omega t \tag{1.4}$$

$$\tau = \tau_0 \sin(\omega t + \delta) \tag{1.5}$$

For analytical purposes, the stress response is decomposed into two waves termed τ' and τ'' (Figure 1.9(B)), which sum to the same amplitude as τ (eqn (1.6)). τ' is defined to be in phase with the strain input, therefore oscillates according to eqn (1.4) ($\sin \omega t$), whereas τ'' oscillates 90° out of phase with the strain ($\cos \omega t$) (eqn (1.6)).

$$\tau = \tau' + \tau'' = \tau_0' \sin \omega t + \tau_0'' \cos \omega t \tag{1.6}$$

These two waves, τ' and τ'', allow for two new dynamic moduli to be defined: the elastic or storage modulus, G', which is in phase with the shear strain;

and the loss modulus, G'', which is out of phase with the shear strain input energy (eqn (1.7) and (1.8), respectively).

$$G' = \tau_0'/\gamma_0 \tag{1.7}$$

$$G'' = \tau_0''/\gamma_0 \tag{1.8}$$

Although these new moduli (G' and G'') may appear to be mathematically abstract constructs, the storage moduli is a measure of the elastic energy retained in the system upon deformation, whereas the loss modulus accounts for the energy dissipated by the sample over each cycle. Thus, by observing how these parameters vary as a function of either temperature or strain for samples before and after fracture, it is possible to build up a detailed picture of the healing behavior of polymer samples.

Rheological healing tests have been successfully conducted on materials using two geometries. The sample may be formed into a 'dogbone' shape which is subject to a linear extension and compression over a defined temperature range: dynamic mechanical thermal analysis (DMTA). Alternatively, the sample may be pressed between a flat base plate and cone shaped top plate which oscillates in a circular manner: cone and plate rheometry (Figure 1.10). Both measurements may be carried out on quantities varying from a few tens of milligrams to many hundreds of grams depending on the equipment used.

In each test, the polymer sample is subjected to the oscillating strain until fracturing. This is observed readily in a DMTA test as the severing of the dogbone sample. Samples tested in the cone and plate geometry are defined as broken when the recorded storage and loss modulus drops to essentially zero. The broken samples are then subject to the designed healing stimulus (if required) and re-tested. Comparison of the storage and loss modulus of the materials before and after healing allowed the healing efficiency of the samples to be assessed. In the case of cone and plate rheometric tests, the broken sample may be left in the instrument for varying periods prior to re-testing, as by the nature of the experiment, the 'fracture faces' are in contact.

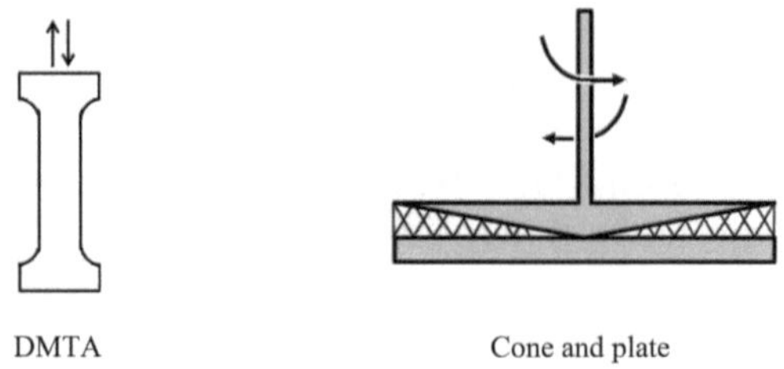

Figure 1.10 Example geometries for samples undergoing rheological healing tests. In the cone and plate geometry, the polymer is placed in the hashed region.

1.6 Observing Healing in Polymeric Systems by Surface Analysis

In addition to the use of tensiometric and rheological assessments of a material's ability to heal either autonomously or whilst exposed to an appropriate stimulus, other analytical techniques have also been employed to study these processes. In particular, monitoring the appearance or profile of the fracture surface by, for example, microscopy can provide excellent visual information on how the polymeric material behaves during the healing process.

1.6.1 Microscopy

1.6.1.1 Optical Microscopy

Studying the surface of a material using optical microscopy[41] in which the image has been amplified *via* a series of optical lenses has proven to be an extremely effective way to assess the material's ability to either heal autonomously or following the application of a suitable stimulus. As shown in Figure 1.11, optical microscopy in conjunction with a variable temperature hot stage enabled the determination of the conditions under which a supramolecular polyurethane can be healed.[42]

1.6.2 Electron Microscopy

In a manner similar to optical microscopy, surfaces of polymers can also be monitored by scanning electron microscopy (SEM)[43] or transmission electron microscopy (TEM).[44] Detailed theoretical treatments of these analytical techniques lies beyond the remit of this text but the reader should be aware

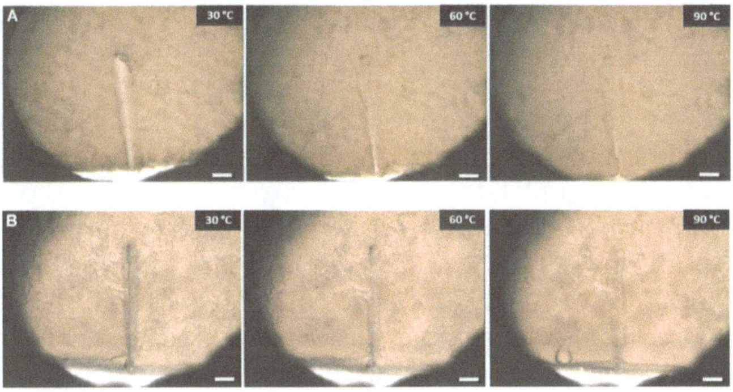

Figure 1.11 Progress of healing monitored by optical microscopy of (A) a supramolecular polyurethane and (B) a supramolecular polyurethane blend. Images were taken using an optical microscope equipped with a hot stage. The scale bar represents 50 μm. Reproduced from ref. 42 with permission from Elsevier, Copyright 2019.

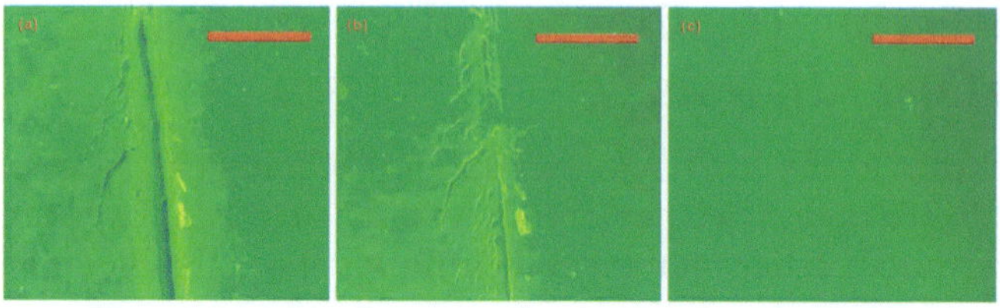

Figure 1.12 False color ESEM images of a healable polymer blend. Micrographs were taken at (a) 25 °C, (b) 100 °C and (c) 200 °C. The images demonstrate both the formation of a homogenous blend and the healability of the polymer at elevated temperatures. The scale bar represents 300 μm. Reproduced from ref. 45 with permission from the Royal Society of Chemistry.

that the enhanced resolution and magnification of the images of the material surfaces affords significantly more information in comparison to optical microscopy with regard to how healing processes evolve. In the past decade, an evolution of SEM has become extremely popular in the analysis of polymer surfaces, this is environmental scanning electron microscopy (ESEM)[ref] – unlike SEM, the sample can be held and monitored under conditions (*i.e.* temperature, humidity) that are more representative of those under which the material will likely be employed in. Shown in Figure 1.12 are a series of images recorded using ESEM of a damaged surface of a supramolecular polymer blend as its temperature was increased – from the false color images the material can be observed to flow into the site of damage in order to recover the uniform surface.[45]

1.7 Measuring Healing by Surface Profiling

The microscopy techniques used to monitor healing outlined in the previous section allow 2D visualization of the facture zone before, during and after healing. However, these methods do not allow the surface profile (deviation from flatness) of the region to be accurately determined. To overcome this deficiency, Hayes and co-workers were able to profile the surface in 3 dimensions over the healing time of the supramolecular polyurethane (PU) they were studying (Figure 1.13).[46]

At the start of the experiment, the PU was damaged by cutting with a razor, and imaged both optically (A1) and by profilometry. During this latter process, the deviation from the mean surface plan is measured to build up a 3D image of the area (B1). The sample was then heated to 37 °C and the two analytical measurements were taken at 30minute intervals (A2 to A6 and B2 to B5). Over this time period the sample can be seen to heal in both the 2D and 3D images. Indeed, after 120 minutes, a transect (red line in B5) taken across the 3D image showed that the surface was essentially flat with no negative deviation in the surface in the damage zone, despite the site of the damage being evidently observable in the optical microscope image.

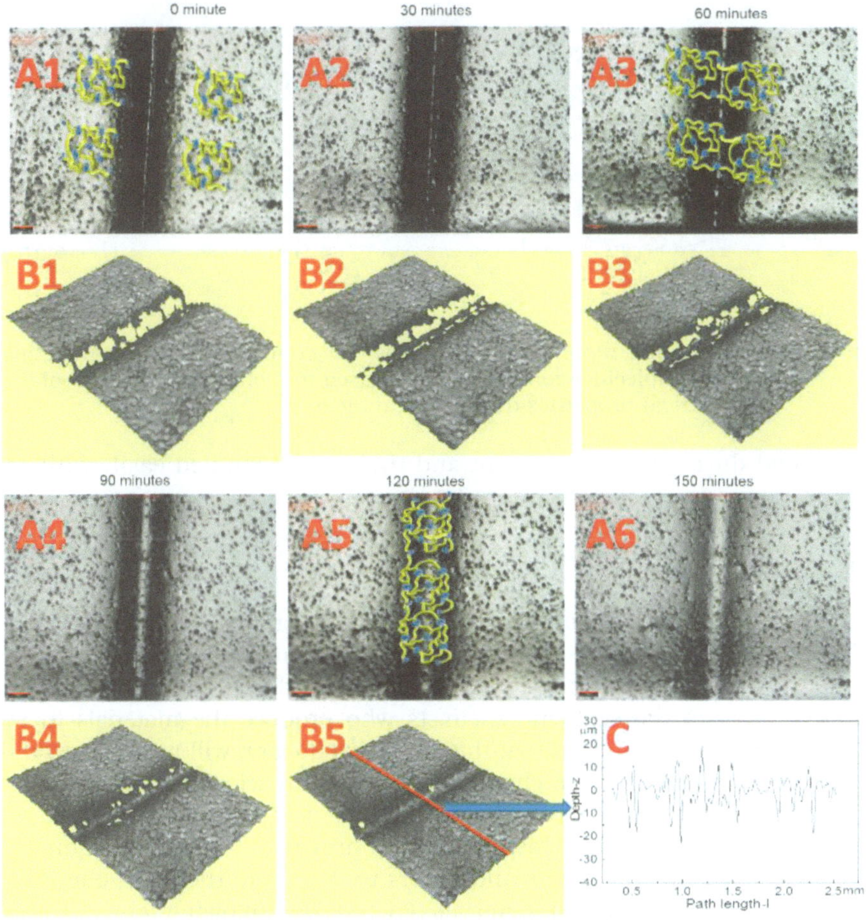

Figure 1.13 The dynamics of the healing of PU at 37 °C by 2D (A1–A6) and 3D (B1–B5) imaging (scale bars 200 μm). The roughness profile (C) is indicated by the red line in B5. Adapted from ref. 46 with permission from the Royal Society of Chemistry.

1.8 Conductivity Measurements to Monitor Healing in Polymeric Samples

Multifunctional polymer composite devices have become a significant focus of research in the field of polymeric materials in the past 20 years. Such composite devices offer the capability to function as sensors or indeed permit property recovery post damage and as such are extremely attractive materials in the development of wearable devices[47] or synthetic 'skins'.[48] The electrical conductivity properties of polymer composites of this type have been used successfully by several groups[49,50] to monitor the healing processes[51] – the ability to conduct current diminishes as the polymer matrix

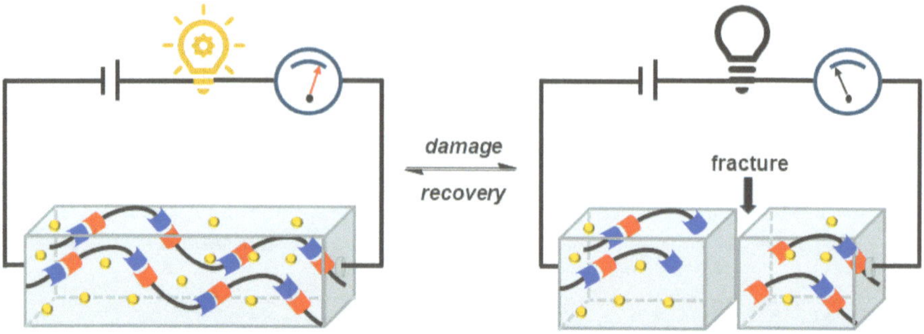

Figure 1.14 Schematic representation of how the conductivity of a conducting supramolecular composite can be used to monitor the recovery of the electrical properties during a healing event.

is disrupted during a damage event and this can be studied easily with an appropriate electrical meter (Figure 1.14).

1.9 Summary

This chapter has provided an overview of the key concepts required to study healable polymeric materials. Many of these techniques, for example impact testing, rheometry and advanced microscopic methods, are far beyond the expertise of most of the synthetic chemists who produce the materials to be analyzed. However, it is hoped that this brief introduction will help narrow the gap in understanding between chemists and materials scientists, further opening up the field of healable materials research to a new generation of scientists.

The following chapters in this monograph provide a reference text on the chemistry of healable polymers and covers to date, and the main chemical strategies used to afford such materials. Each of the following four chapters provides a critical assessment of the main developments on healable polymer systems and, where appropriate, concludes with examples of where these materials reach (or are expected to reach) market:

1.9.1 Chapter 2: Encapsulation-based Self-healing Polymers and Composites

This chapter provides detailed coverage of the development of the encapsulated monomer approach to healable polymer composites, and also includes a summary of mechanistic aspects with specific reference to key examples from the literature.

1.9.2 Chapter 3: Reversible Covalent Bond Formation as a Strategy for Healable Polymer Networks

Reversible covalent bond formation processes such as Diels–Alder cycloadditions have been utilized successfully to access healable polymer networks.

This chapter will examine the range of reversible covalent bond formation processes that have been used to this end.

1.9.3 Chapter 4: Healable Supramolecular Polymeric Materials

Recent developments in healable polymer and gel systems have employed key concepts from the field of supramolecular chemistry to afford exciting materials that can be repaired with high efficiency and reproducibly. Use of non-covalent interactions such as electrostatic forces of attraction, hydrogen bonding, metal–ligand binding and π–π aromatic stacking have been used to assemble polymeric networks built from relatively low molecular weight components – the hypothesis underpinning the healing characteristics of these materials is that disruption of weak non-covalent interactions facilitates repair prior to fracture of the covalent linkages in the oligomers.

1.9.4 Chapter 5: Thermodynamics of Self-healing Reactions and their Application in Polymeric Materials

This chapter will provide detailed coverage of the thermodynamic considerations of healable materials. The role of the entropic and enthalpic contributions to the Gibbs free energy during self-healing events will be investigated, with a specific study on how the configurational changes of the polymers and chemical reactions within the material contribute to driving healing in bulk materials.

References

1. The are several excellent popular accounts concerning the concept of the Plastics Age for example: (i) *The Plastics Age: From Bakelite to Beanbags and Beyond*, Professor Penny Sparke, Overlook Press.
2. For an essay concerning Staudinger's life and achievemccents see: R. Mülhaupt, *Angew. Chem., Int. Ed.*, 2004, **43**, 1054.
3. For a contemporary review of his own work: W. H. Carothers, *Chem. Rev.*, 1931, **8**, 353.
4. Source: Statistica: https://www.statista.com/statistics/1307217/polymer-sales-by-segment-european-union/ (Retrieved July 2023).
5. J. Datta and P. Kopczynska, *Crit. Rev. Environ. Sci. Technol.*, 2016, **46**, 905.
6. T. Thiounn and R. C. Smith, *J. Polym. Sci.*, 2020, **58**, 1347.
7. A. Soroudi and I. Jakubowicz, *Euro. Polym. J.*, 2013, **49**, 2839.
8. K. M. Zia, H. N. Bhatti and A. I. Bhatti, *React. Funct. Polym.*, 2007, **67**, 675.
9. Several excellent books covering healable and adaptive materials have been published (see also ref. 10 and 11): *Self Healing Materials: an Alternative approach to 20 centuries of materials science*, ed. S. van der Zwaag, Springer, Dordrecht, 2007.

10. *Self Healing Materials: Fundamentals, Design Strategies and Applications*, ed. S. K. Ghosh, J. W. Wiley and sons Ltd., Weinheim, 2009.
11. *Adaptive Structures: engineering applications*, ed. D. Wagg, I. Bond, P. Weaver and M. Friswell, J. W. Wiley and sons Ltd., Chichester, 2007.
12. Several excellent reviews covering healable polymers have been published (references 12 to 18): R. P. Wool, *Soft Matter*, 2008, **4**, 400.
13. S. D. Bergman and F. Wudl, *J. Mater. Chem.*, 2008, **18**, 41.
14. Y. C. Yuan, T. Yin, M. Z. Rong and M. Q. Zhang, *Polym. Lett.*, 2008, **2**, 238.
15. S. Burattini, B. W. Greenland, D. Chappell, H. M. Colquhoun and W. Hayes, *Chem. Soc. Rev.*, 2010, **39**, 1973.
16. T. C. Mauldin and M. R. Kessler, *Inter. Mater. Rev.*, 2010, **55**, 317.
17. E. B. Murphy and F. Wudl, *Prog. Polym. Sci.*, 2010, **35**, 223.
18. J. A. Syrett, R. C. Becer and D. M. Haddleton, *Polym. Chem.*, 2010, **1**, 978.
19. W. Denissen, J. M. Winne and F. E. Du Prez, *Chem. Sci.*, 2016, **7**, 30.
20. A. Liguori and M. Hakkaraninen, *Macromol. Rapid Commun.*, 2022, **43**, 2100816.
21. L. R. Hart, T. S. Babra, J. L. Harris, B. W. Greenland and W. Hayes, *Polym. Chem.*, 2013, **4**, 4860.
22. A. D. O'Donnell, S. Salimi, L. R. Hart, T. S. Babra, B. W. Greenland and W. Hayes, *React. Funct. Polym.*, 2022, **172**, 105209.
23. Y. Yanagisawa, Y. Nan, K. Okuro and T. Aida, *Science*, 2018, **359**, 72.
24. Source: https://www.marketresearchfuture.com/reports/self-healing-polymer-market-4135 (accessed September 2023).
25. See, for example: A. J. Capel, R. P. Rimington, M. P. Lewis and S. D. R. Christie, *Nat. Rev. Chem.*, 2018, **2**, 422.
26. G. Castro-Hermosa, G. Lucarelli, V. Manca, G. Farinola and T. M. Brown, *Adv. Funct. Mater.*, 2019, **29**, 1806798.
27. J. Persad and S. Rocke, *Results Eng.*, 2022, **16**, 100730.
28. Y. Bai, L. Wilbraham, B. J. Slater, M. A. Zwijnenburg, R. S. Sprick and A. I. Copper, *J. Am. Chem. Soc.*, 2019, **141**, 9063.
29. W. Zhao, P. Y. Yan, B. Y. Li, M. Bahri, L. J. Liu, X. Zhou, R. Clowes, N. D. Browning, Y. Wu, J. W. Ward and A. I. Cooper, *J. Am. Chem. Soc.*, 2022, **144**, 9902.
30. A. J. Gormley and W. A. Webb, *Nat. Rev. Mater.*, 2021, **6**, 642.
31. M. M. Cencer, J. S. Moore and R. S. Assary, *Polym. Int.*, 2022, **71**, 537.
32. P. C. Xu, H. M. Chen, M. J. Li and W. C. Lu, *Adv. Theory. Simul.*, 2022, **5**, 202100565.
33. X. B. Li, P. M. Maffettine, Y. Che, T. Liu, L. J. Chen and A. I. Copper, *Chem. Sci.*, 2021, **12**, 10742.
34. Source: https://www.nissa n-global.com/EN/INNOVATION/TECHNOLOGY/ARCHIVE/SCRATCH/ (accessed September 2023).
35. Source: www.bayercoatings.de (accessed September 2023).
36. Source: https://www.marketsandmarkets.com/Market-Reports/self-healing-coatings-market-96446404.html (accessed September 2023).
37. For an excellent review of CB testing with respect to healing materials see: N. Brown, *J. Strain. Anal.*, 2011, **46**, 167.

38. *Introduction to physical metallurgy*, ed. S. H. Avner, McGraw-Hill, New York, 2nd edn, 1974.
39. *Metallurgy for Engineers*, ed. E. C. Rollason, Edward Arnold, London, 4th edn, 1973.
40. *Rheology: Principles, Measurements and Applications*, C. W. Macosko, Wiley-VCH Inc, New York, 1994.
41. W. Wang, *Chem. Soc. Rev.*, 2018, **47**, 2485.
42. S. Salimi, L. R. Hart, A. Feula, D. Hermida-Merino, A. B. R. Touré, E. A. Kabova, L. Ruiz-Cantu, D. J. Irvine, R. Wildman, K. Shankland and W. Hayes, *Eur. Polym. J.*, 2019, **118**, 88.
43. L. E. Franken, K. Grünewald, E. J. Boekema and M. C. A. Stuart, *Small*, 2020, **16**, 1906198.
44. Foundations of Environmental Scanning Electron Microscopy: G. D. Danilatos, *Adv. Electron. Electron Phys.*, 1988, **71**, 109.
45. L. R. Hart, J. H. Hunter, N. A. Nguyen, J. L. Harries, B. W. Greenland, M. E. Mackay, H. M. Colquhoun and W. Hayes, *Polym. Chem.*, 2014, **5**, 3680.
46. A. Feula, X. Tang, I. Giannakopoulos, A. M. Chippindale, I. W. Hamley, F. Grecco, C. P. Buckley, C. R. Siviour and W. Hayes, *Chem. Sci.*, 2016, **7**, 4291.
47. S. Lage-Rivera, A. Ares-Pernas and M.-J. Abad, *Inter. J. Energy Res.*, 2022, **46**, 10475.
48. D. E. Snyder and E. D. Sapper, *J. Compos. Sci.*, 2022, **6**, 141.
49. H.-J. Wang, Y.-Z. Chu, C.-K. Chen, Y.-S. Liao and M.-Y. Yeh, *RSC Adv.*, 2021, **11**, 6620.
50. J. Liu, C. S. Y. Tan, Z. Yu, N. Li, C. Abell and O. A. Scherman, *Adv. Mater.*, 2017, **29**, 1605325.
51. For a review of supramolecular healable composites see: R. Vaiyapuri, B. W. Greenland, H. M. Colquhoun, J. M. Elliott and W. Hayes, *Polym. Int.*, 2014, **63**, 933.

CHAPTER 2

Encapsulation-based Self-healing Polymers and Composites

MICHAEL W. KELLER

Department of Mechanical Engineering, The University of Tulsa, 800 S. Tucker Drive, Tulsa, OK 74104, USA
Email: mwkeller@utulsa.edu

2.1 Introduction

Researchers demonstrated the repair of polymeric materials *via* the addition of solvents or adhesives in the 1980s.[1] However, these approaches lacked the ability to initiate and complete the repair automatically. Some form of outside intervention, either through the addition of solvent or through the application of heat or pressure, was required. Autonomy was introduced when White *et al.* demonstrated the first microcapsule-based self-healing material.[2] The introduction of microcapsules enabled the synthesis of a truly self-healing material that initiated and completed self-repair without any outside intervention. The capsules provided both stable storage and reliable release and were the transformative advance that made autonomic healing possible. Since that initial demonstration, the use of encapsulation as the basis for introducing self-healing functionality has expanded rapidly.

The specific details of any given encapsulation-based self-healing material may differ, but the basic mechanism for the incorporation and activation of the functionality remain the same. An encapsulated healing agent is embedded into a host matrix along with either a catalyst or a cross-linker phase. Damage, typically a crack, then propagates through the material and

Polymer Chemistry Series No. 41
Healable Polymer Systems: Fundamentals, Synthesis and Applications, 2nd Edition
Edited by Barnaby W. Greenland and Wayne Hayes

Published by the Royal Society of Chemistry, www.rsc.org

ruptures the embedded capsules. The released healing agent is pulled into the damage by capillary action where contact with a catalyst or a released cross-linker phase initiates polymerization. This initiator/activator phase has been both a solid that is dissolved by the released healing agent and a liquid that requires some amount of diffusive mixing. After contact with the catalyst/cross-linker, the healing agent polymerizes and bonds the crack faces together. The new adhesive bond is what restores the original mechanical properties of the host matrix. Schematically, this process of damage, response, and healing is shown in Figure 2.1.

Critical to the success of encapsulation-based self-healing materials is the interaction of the damage/fracture with the embedded capsules. One of the advantages of the microcapsule approach presented by White and coworkers is that cracks are attracted to the midplane of a microcapsule when the modulus of the capsule is less than that of the surrounding matrix.[2] An Eshlby-Mura equivalent inclusion model showed that the interaction of the stress-field around an off-center-crack and an inclusion would tend to drive the crack to the centerline of the microcapsule.[2] This prediction was confirmed by experiment.

In this chapter the synthesis and characterization of microcapsule-based self-healing polymers will be discussed. The microencapsulation processes that have been developed for the encapsulation of healing agents and the relevant healing chemistries will be covered. Subsequently, a review of the

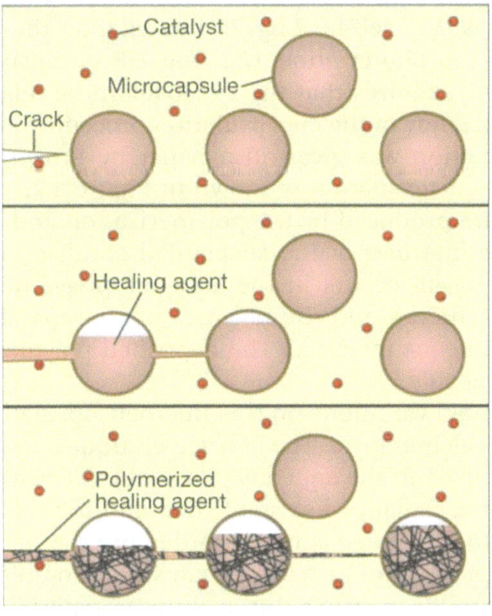

Figure 2.1 Schematic representation of a micro-capsule-based self-healing material. Reproduced from ref. 2 with permission from Springer Nature, Copyright 2001.

testing and performance of self-healing polymers and composites is presented. Finally, a selection of applications of healable polymeric materials containing microcapsules is presented to provide the reader with an indication of the potential pathway towards commercialization of these fascinating materials.

2.2 Microencapsulation

The encapsulated phase is the critical component of this class of self-healing materials. Capsules are the self-healing mechanism, serving both as storage and trigger of the healing response. Therefore, significant literature has been developed investigating the encapsulation of a wide variety of chemically active agents intended for use in self-healing materials. These encapsulation approaches draw on the related literature that has been devoted to encapsulation of materials ranging from fragrance to pharmaceuticals. Several comprehensive reviews of encapsulation exist, see for example ref. 3 and 4, but a few important techniques that are frequently used within the context of self-healing materials will be briefly described.

2.2.1 *In Situ* Poly(Urea-formaldehyde) Encapsulation

Encapsulation in a poly(urea formaldehyde) or melamine formaldehyde shell has proven to be one of the most versatile and widely used encapsulation approaches within self-healing. In this chapter these systems will be referred to using the abbreviation UF. The UF encapsulation process is emulsion-based and requires that the encapsulant be relatively immiscible in water. A detailed study of the encapsulation procedure and analysis of the microcapsule formation was given in a paper by Brown *et al.*[5] The basic procedure for this encapsulation is shown in Figure 2.2.

Microcapsules are produced by the polymerization and deposition of the UF polymer at the interface of the suspended emulsion droplet. Polymerization of the shell-wall occurs in the aqueous phase until a critical molecular weight is achieved and the polymer phase-separates, depositing at the encapsulant/aqueous interface. The deposited polymer forms the shell-wall of the microcapsule.

There are two broad variations on this microencapsulation approach that have been used in the literature. The *in situ* technique uses direct addition of urea and formaldehyde to the aqueous phase of the encapsulation bath. An emulsion of the encapsulant was established and then urea and formaldehyde were added to the aqueous phase. Adjusting the pH to slightly acidic and heating the encapsulation bath initiates and completes the encapsulation. Capsules produced using this technique typically have thin shell-walls, on the order of a few hundred nanometers, with a relatively thick porous layer of UF nanoparticles attached to the surface.[5] However, if the encapsulant to wall-material ratio is high enough, nearly the entire mass of

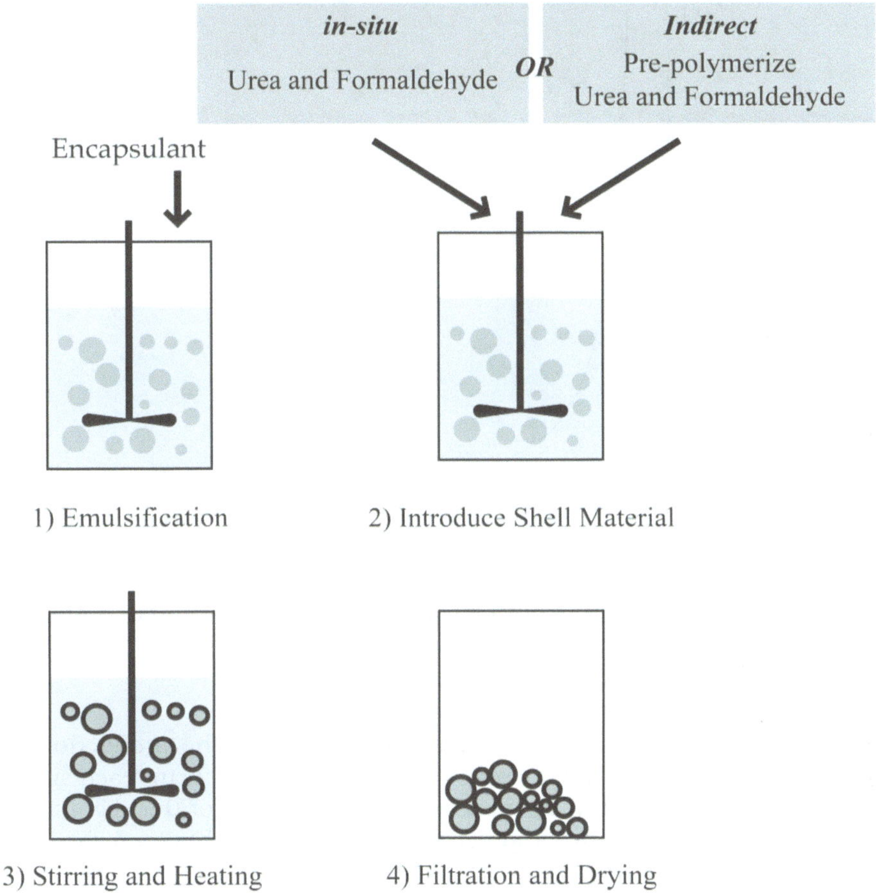

Figure 2.2 Schematic of the UF encapsulation procedure.

shell-wall precursors can be consumed to produce the shell wall resulting in microcapsules with surfaces that are relatively smooth.

An indirect technique can also be used to produce UF microcapsules. In this approach, a prepolymerization of urea, melamine, or phenol and formaldehyde is performed in a separate reaction vessel. The prepolymer is then added to the emulsified core material at the point where the urea and formaldehyde would normally be added in the *in situ* process. Encapsulation proceeds as it would in the *in situ* process. Capsules formed using the indirect procedure can have much thicker shell walls, up to several microns.[6]

Emulsions for both encapsulation approaches are typically produced using physical stirring. The mixer blades induce a strong shear flow that breaks the encapsulant into small droplets. Capsule diameter is then directly relatable to stirring speed, as an example Figure 2.3 shows the diameter of dicyclopentadiene-filled (DCPD) microcapsules produced using the *in situ* technique.

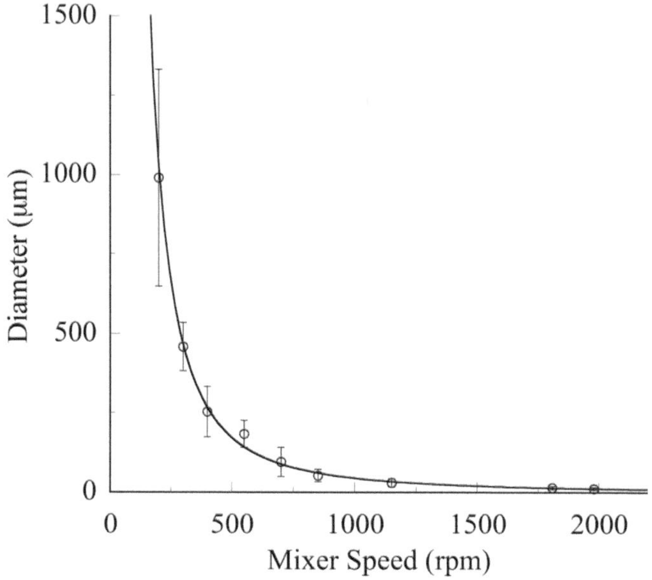

Figure 2.3 Size of DCPD-containing microcapsules as a function of stirrer speed. Reproduced from ref. 6 with permission from Taylor & Francis, Copyright 2010.

While the mechanical agitation approach allows for the production of capsules as small as 20 to 30 microns in diameter, the ability to reach single micron and below capsules is generally beyond the reach of this technique. Therefore, several alternative approaches to synthesize submicron capsules have been presented in the literature. The first demonstration of submicron-sized microcapsules utilized a combined ultrasonic dispersion and stirring to generate an emulsion with nanometer-sized droplets.[7] Capsules were synthesized by irradiating a small-volume capsule batch with ultrasound and then stirring to maintain the dispersion during the standard *in situ* UF encapsulation procedure. This process was capable of producing capsules between one micron and several hundred-nanometers in diameter.

Other researchers have used the Shirasu Porous Glass method to generate an emulsion with a narrow size distribution.[6] Emulsions are produced by forcing an immiscible liquid through the membrane into a host liquid. These systems have demonstrated the capability to produce nearly monodisperse liquid and gas emulsions. In an encapsulation setting, the membrane is used to form the droplets in the emulsion for an indirect UF encapsulation. One of the novel features of the reported procedure was the construction of a rotating membrane that served as both the emulsion source and the mechanical agitator that maintained the emulsion. While this approach yielded capsules that were within the capability of mechanical stirring, *ca.* 40 microns, the distribution of diameters was much narrower when compared to mechanically emulsified encapsulation procedures.

In addition to direct formation of microcapsules filled with a core of healing agent, the UF encapsulation procedure has also been successful in manufacturing capsules through a diffusion-based process. Encapsulation through this method is performed by first producing either hollow or solvent-filled capsules. As mentioned above, the indirect UF procedure produces capsules with relatively thick shell walls. These capsules are, therefore, far more stable to buckling collapse when compared to the capsules produced by the *in situ* technique and can be used to encapsulate air or gas bubbles entrained in the encapsulation bath. These bubbles can be produced by including a blowing agent in the encapsulation bath or by aggressively stirring an encapsulation bath to entrain air. The hollow capsules can then be filled using a diffusion approach. Capsules are dispersed in a liquid of interest, which diffuses inward through the capsule shell-wall, eventually filling as osmotic pressure equalizes. This approach has proven useful for encapsulating highly reactive or water sensitive materials such as boron trifluoride diethyl etherate.[8]

One continuing concern for the microcapsule-based approach is the stability of the capsules. The diffusive mobility of the core material across a thin polymer shell is a benefit for manufacturing capsules with core materials that cannot be encapsulated, but also presents a problem when storing the capsules in a dry form. The stability of the capsules for a small range of core materials was investigated in a paper by Nesterova, Dam-Johansen, and Kiil.[9] In this paper both the UF and melamine–formaldehyde encapsulations were performed with an epoxy resin as the core material. Capsules were found to be unstable after long storage in air, approximately 7 weeks. To address these stability issues, researchers have developed two-step encapsulation procedures that synthesize capsules with two shell-walls. One of the first reports was the silica coating of an existing UF-walled, submicron capsule containing dicyclopentadiene.[10] A cross-section of the capsule is shown in Figure 2.4.

Capsules were coated using a fluoride catalyzed condensation reaction using tetra orthosilicate as the starting material. An additional benefit, aside from the added stability, was an improved resistance to agglomeration, which made smaller capsules easier to handle.

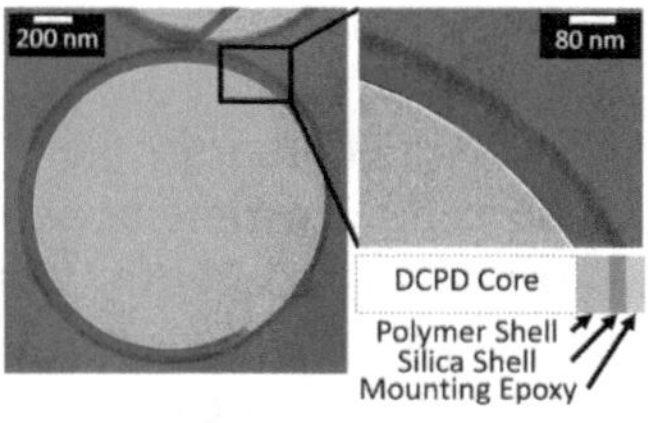

Figure 2.4 TEM image of a microtomed silica-protected microcapsule containing DCPD. Reproduced from ref. 10 with permission from John Wiley & Sons, Copyright 2011 Wiley-VCH Verlag GmbH & Co. KGaA, Weinheim.

2.2.2 Interfacial Polymerization

The second most popular encapsulation procedure for self-healing materials is the interfacial polymerization approach. This is a fundamentally different process when compared to the *in situ* urea or melamine-formaldehyde process. Instead of a polymerization reaction that proceeds mostly within the aqueous phase of the encapsulation bath, the polymerization reaction in the interfacial encapsulation procedure occurs at the encapsulant/water interface building the shell wall by forming a film that thickens with time.

Interfacial polymerizations are typically step-growth reactions where one monomer is hydrophilic and the second monomer is hydrophobic. An interfacial polymerization is classically demonstrated in the reaction of a diamine in an aqueous solution and a sebacoyl chloride in a non-polar solvent. This arrangement is referred to as the "nylon rope trick," as a thin cord of nylon polymer can be drawn up from the film that forms at the interface of the aqueous and organic phases. In encapsulation, this is exploited by dissolving a hydrophobic prepolymer or monomer in the intended encapsulant. A co-solvent is sometimes required. This solution is then emulsified as with the UF system. An aqueous monomer is then added to the encapsulation bath and the polymerization proceeds at the interface between the encapsulate and the water, building the polymer around the droplet and forming the shell-wall.

Interfacial encapsulations based on polyurethane chemistry have been the most common for self-healing materials. In a typical encapsulation, a prepolymer derived from an aromatic diisocyanate, typically toluene diisocyanate (TDI) or methyldiphenyl diisocyanate (MDI), is dissolved in the encapsulant phase. A co-solvent is frequently required to ensure that the prepolymer and encapsulant are miscible. The resulting solution is then emulsified in an aqueous bath that contains an emulsion stabilizer. Gum arabic is commonly used, but other stabilizers have also been successful.[11,12] After the emulsion has stabilized, a cross-linker or chain extender is added to the aqueous phase. In the case of polyurethanes, diols such as ethylene diol and butanediol are typically used. Polymerization occurs at the interface between the encapsulant and the aqueous phase forming the shell wall (Figure 2.5). The microcapsules can have relatively thick shell walls, on the order of a few microns. However, unlike the UF system, shell wall thickness can be controlled with reasonable fidelity by varying the reaction time or mass of the available shell wall precursors.

This approach has been used for a variety of healing agents, ranging from an organotin catalyst[11] to amine curing agents.[13] One of the more interesting approaches was to utilize the reaction rate difference between aromatic diisocyanates and aliphatic diisocyanates to encapsulate a reactive isocyanate core. In the first demonstration of this encapsulation procedure, a TDI-based prepolymer was utilized to encapsulate isophorone diisocyanate.

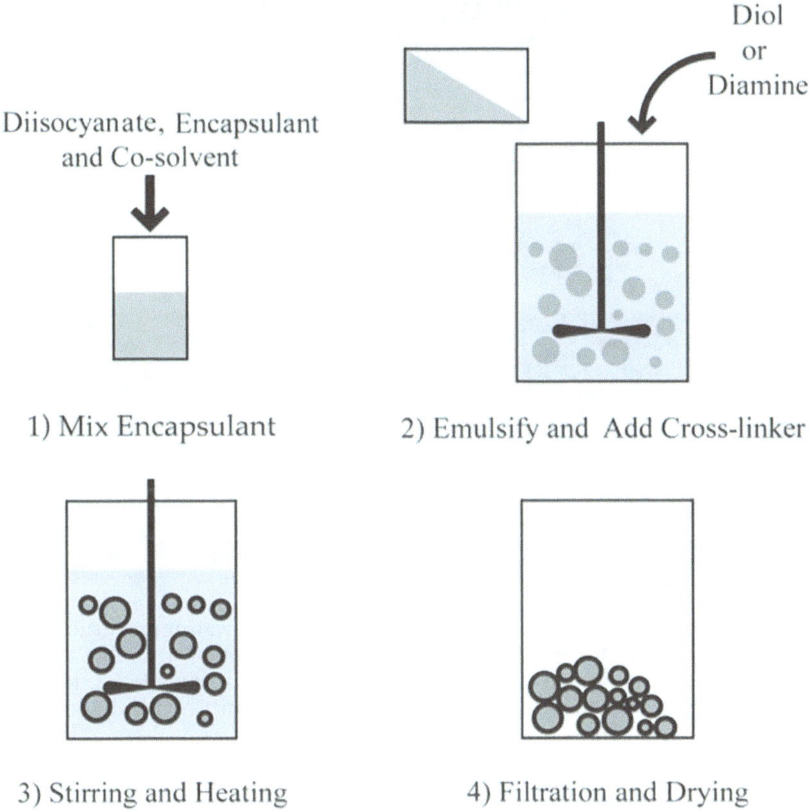

Figure 2.5 Schematic of the polyurethane interfacial encapsulation procedure.

The three order-of-magnitude differences in reaction rates between the aliphatic IPDI and the aromatic TDI led to preferential TDI consumption and produced microcapsules containing a reactive isophorone core.[14] Building on this approach, Yang and co-workers utilized a similar approach with an MDI-based prepolymer to encapsulate a hexamethylene diisocyanate core material.[15]

An interesting combination of the two microencapsulation procedures described above was used to produce a highly-stable, double-walled microcapsule formed of an inner polyurethane (PU) shell microcapsule with an outer urea-formaldehyde capsule in one step, unlike the two step encapsulations described above.[16] A SEM and AFM phase image of the double capsule wall is shown in Figure 2.6. The single step production was accomplished by simply running both the *in situ* UF and interfacial polymerizations at the same time. Kinetic differences in the two polymerizations led to the formation of a PU inner shell and a UF outer shell.

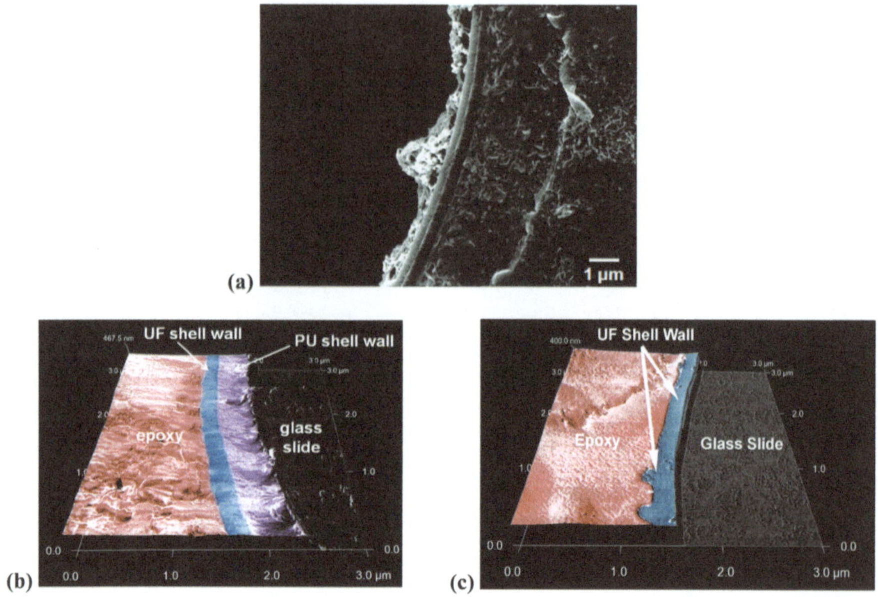

Figure 2.6 A SEM image (a), showing the double-shell wall microstructure of the capsule. Images (b) and (c) are AFM images of a double wall microcapsule, image (b), and a single walled microcapsule (c). Reproduced from ref. 16 with permission from American Chemical Society, Copyright 2010.

These double-wall microcapsules demonstrated exceptional stability, showing retention of the core material after one year of storage with little or no degradation. Furthermore, these capsules also demonstrated significantly better stability when exposed to high temperatures, as might be experienced during a composite cure cycle.

While two-layer capsules were successful in surviving high cure temperatures, these capsules were improved by Kang *et al.*[17] *via* the addition of a third layer. By overcoating a double-walled PU/UF capsules with a polymerized dopamine, enhanced thermal stability was achieved. The onset of mass loss, which indicates the loss of core material, occurred at 245 °C, higher than in double layered capsules. Further characterization of these microcapsules under isothermal conditions confirmed negligible mass loss at 180 °C for 3 hours. Uncoated PU/UF microcapsules experienced over 94% mass loss when exposed to the same conditions. Solvent and pH resistance of these microcapsules was also much improved, with the triple wall PDA-coated microcapsules performing better than uncoated microcapsules.

Since these initial demonstrations, other multilayer microcapsules have also been synthesized. Yi *et al.* encapsulated isophorone diisocyanate *via in situ* polymerization of melamine formaldehyde prepolymer in a Pickering emulsion stabilized by lignins.[18] The lignins performed two functions, the first was enhancing the emulsion stability of the core material and the second was to form an outer layer on the microcapsules upon completion of the melamine formaldehyde reaction. By incorporating both mechanisms, a very thick shell, on the order of 5 μm was formed. Themal stability was very good, with the onset of mass loss occurring at 204 °C during the temperature ramp, when a core material with a boiling temperature of 132 °C was encapsulated.

2.2.3 Hollow Fiber Encapsulation

An alternative to the broadly chemical microencapsulation schemes described above is the use of hollow fibers as the container for a given healing agent. These fibers can be as simple as a heat-sealed glass pipette,[19] but the most successful systems use fibers produced from an advanced fiber-pulling system.[20] This approach takes a fiber preform of the desired fiber geometry and then draws out the fiber that retains the initial shape of the preform, but at much reduced diameter, shown schematically in Figure 2.7.

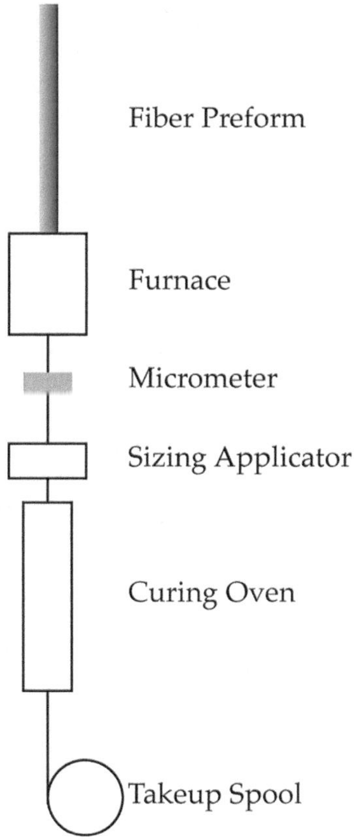

Figure 2.7 Schematic of the fiber pulling process. The fiber preform is loaded into a furnace that heats the material and allows a filament to be pulled to a smaller diameter.

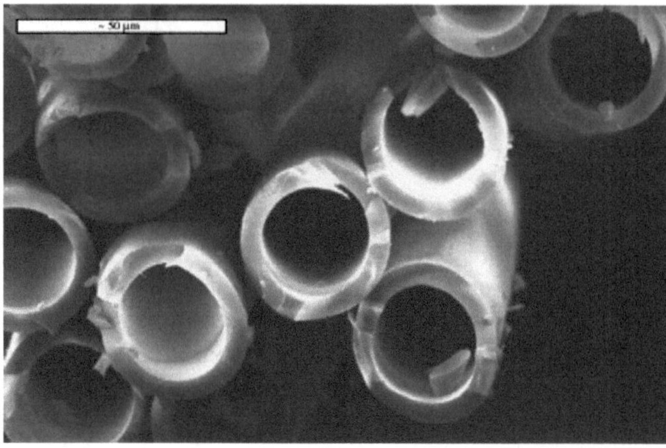

Figure 2.8 Hollow glass fibers produced using the fiber drawing process. Photo courtesy of Professor Ian Bond of the University of Bristol.

Nearly any fiber shape can be generated using this approach, but hollow fibers, such as those shown in Figure 2.8 are employed in the same manner as microcapsules. Fibers are simply incorporated into a composite assembly where they serve as the repositories for healing agents. Filling of the fibers is completed through a vacuum-assisted infiltration process that pulls the desired core material into the embedded hollow fibers.[20] At the moment, these fibers are not prefilled prior to the fabrication of a composite material.

In addition to the glass-fibers, multi-component polymeric fibers have also been described in the literature.[21] To manufacture these multi-component fibers, an emulsion of the target encapsulant and the fiber material was made and then charged into a fiber spinning apparatus. The reported system utilized dichlorobenzene (DCB) as the encapsulant, and the fibers used were alginate. The fiber was then drawn out and the emulsified droplets became encapsulated within the fiber as shown in Figure 2.9. While DCB has been utilized as a self-healing chemistry system by other researchers,[22] the fibers were not used within a self-healing material.

Finally, an electrospinning-based encapsulation scheme has also been demonstrated for the encapsulation of potential healing agents.[23] Electrospinning is a modified fiber-spinning process that combines standard polymer jet spinning with a high-voltage electrical field. The electrostatic forces generated by the applied electric field creates a very narrow jet that projects from the polymer solution, enabling the formation of thin polymer fiber. Final fiber diameters can be on the order of a few hundred nanometers. Encapsulation by electrospinning uses a coaxial approach where a set of nested spinning needles is used to create a coaxial polymer jet with the shell material on the exterior and the target encapsulant in the inner stream. In the case of liquid inner cores, a beaded morphology is typically produced. Indeed, Park and Braun[23] have utilized successfully a coaxial electrospinning

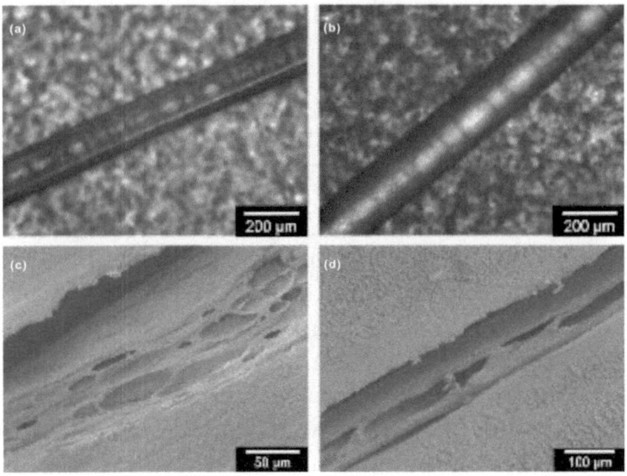

Figure 2.9 Microscope images of a fiber containing compartmentalized droplets in (a) and (b) and a microtome SEM showing the hollows in a mounted and extracted fiber (c) and d). Reproduced from ref. 21 with permission from Elsevier, Copyright 2012.

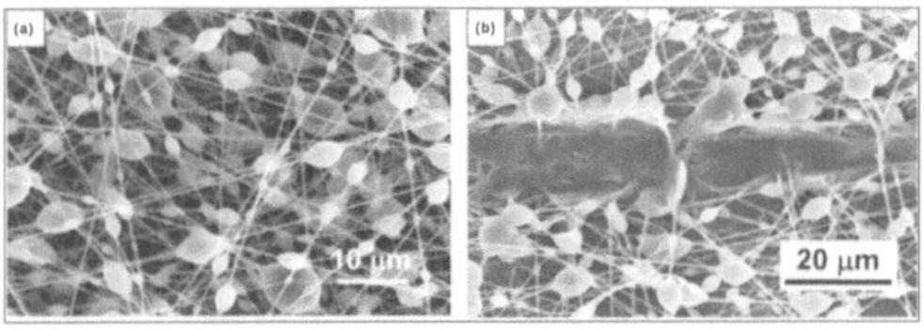

Figure 2.10 Beaded fibers containing a poly(dimethyl siloxane) healing agent. Fibers were formed from a poly(vinylpyrrolidine) polymer shell. Reproduced from ref. 23 with permission from John Wiley & Sons, Copyright 2010 Wiley-VCH Verlag GmbH & Co. KGaA, Weinheim.

process to produce fibers with a beaded morphology that contained a poly(dimethylsiloxane) healing agent as shown in Figure 2.10.

In addition to the encapsulation of siloxane, DCPD and isophorone diisocyanate (IPDI) have also been encapsulated by electrospinning.[24] Sinha-Ray *et al.* utilized a polyacrylonitrile shell to encapsulate the target materials, but proceeded in a slightly different manner than previous researchers. While fibers containing DCPD and IPDI were encapsulated using a coaxial approach as described above, fibers were also fabricated using an emulsion technique and a solution blowing process.[24] In emulsion electrospinning, an emulsion of core and shell material is premade prior to electrospinning. This results in a randomly distributed core material housed within the electrospun fiber, similar to the microstructure shown in Figure 2.10. In solution blowing, a high-speed air jet is directed coaxially around a spinning needle and induces a bending instability similar to that of electrospinning. A fiber is synthesized without the requirement of electrical conductivity needed for successful electrospinning. None of the fibers produced in this study were utilized to produce a self-healing material, but the desired reactivity was confirmed.

2.2.4 Microfluidic Encapsulation

As an alternative to the bulk emulsion approaches described above, microfluidics can be successfully used to produce microcapsules. These approaches are typically based on a microfluidic flow control device to create a multiple emulsion that envelopes a droplet of core fluid in a shell wall material.[25,26] This technique has been widely adopted for a range applications including drug delivery, carbon capture, and self-healing or sensing. Several variations on devices exist for the fabrication of microcapsules *via* microfluidic systems. One of the original approaches was a flow focusing device first demonstrated by Utada *et al.* This system, shown in Figure 2.11, consisted of nested capillary tubes that allowed for the generation of monodispere droplets or monodisperse droplets enveloped in an outer shell fluid, which could later be hardened to form microscapules. In this process, three fluids are used. The middle fluid is not miscible with either the inner fluid or the outer fluid. The inner fluid is pumped out of a nozzle that is surrounded by a middle fluid flowing in the same direction. The droplets that form are a sphere of inner fluid coated in a layer of the middle fluid. These droplets are then carried out of the device by the outer fluid.

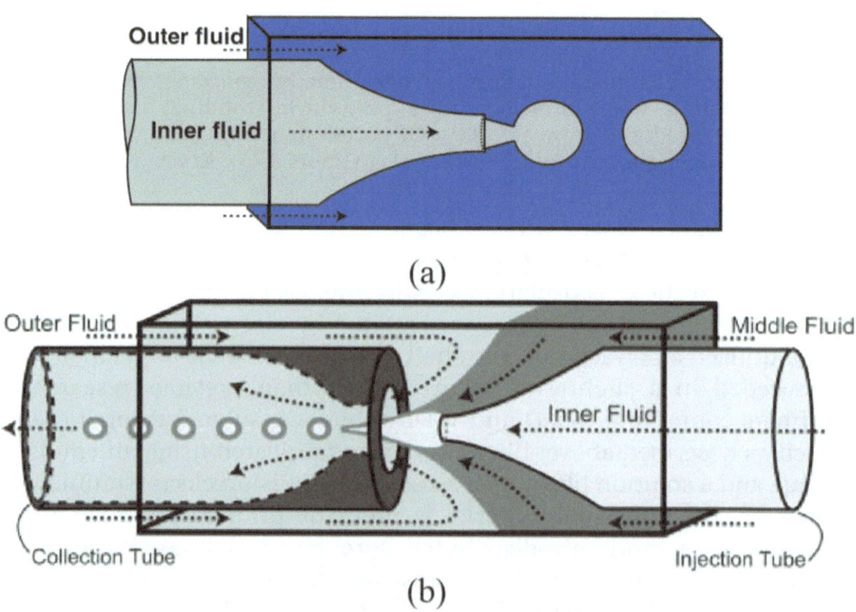

Figure 2.11 Schematics of (a) single emulsion and (b) double emulsion flow-focusing-based microfluidic devices. Reproduced from ref. 26 with permission from AAAS, Copyright 2005.

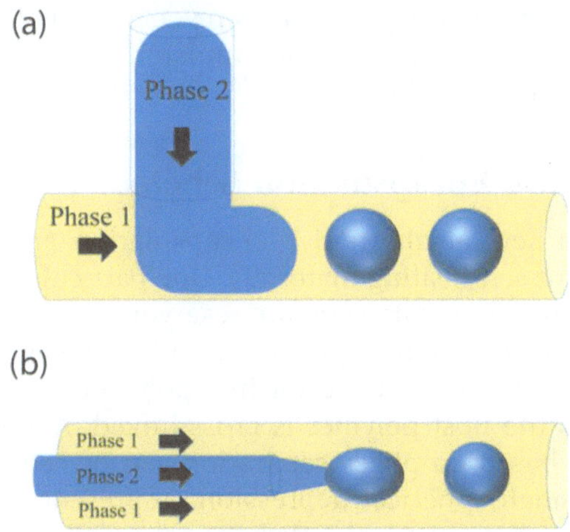

Figure 2.12 Microfluidic junction geometries for (a) cross-flow and (b) co-flow that are commonly used for the production of mono-disperse liquid droplets. Adapted from ref. 27 with permission from John Wiley & Sons, Copyright 2020 Wiley-VCH Verlag GmbH & Co. KGaA, Weinheim.

A schematic of this process is seen in Figure 2.11. For making microcapsules with a solid shell and liquid core, a pre-polymer is used for the middle fluid, and the liquid core material is the inner fluid. The outer fluid is a mixture of surfactants to help stabilize the emulsion. Once the droplets are formed, they are collected and cured *via* various techniques, with photopolymerization being the most common.

In addition to the flow-focusing methods, there are two additional commonly used geometries. These include the cross flow approach and the co-flow approach, shown in Figure 2.12.

These geometries, along with the flow focusing device shown in Figure 2.10(a), represent the basic building blocks of nearly all microfluidic-based encapsulation approaches.[25,28] One challenge that microfluidic encapsulation faces is the clogging of these sensitive microfluidic systems.[29] This clogging is caused by many factors, but includes unwanted curing of the shell material stream or the accumulation and jamming of partially cured microcapsules in the collection tube. In addition to instabilities in the flow that may cause unwanted core fluid intrusion into the carrier fluid. An alternative is to replace the flow-focusing approaches with mechanical vibration aided systems.[30] Which can eliminate the need for a carrier fluid and the associated clogging potential downstream of the droplet production system.

One of the critical advantages of microfluidic encapsulation is the broad flexibility of shell materials and core materials. Shell materials include synthetic polymers, natural polymers, and inorganic materials, where cores can be aqueous, oil-based, or solid. The ability to easily form capsules with aqueous cores is particularly powerful as reverse emulsions are complicated to form and only a few reports of emulsion-based encapsulation with aqueous cores exist in the literature.[31–33]

2.2.5 Alternative Encapsulation Schemes

There have been several alternative encapsulation approaches that have been successful for self-healing materials. One particularly novel idea was to use phase separation to create separate reservoirs of healing agent within a polymer.[11] By choosing the host polymer and healing agent appropriately, a blended mixture of the two will result in a polymer with a phase separated healing agent, as the host polymer is cross-linked. A SEM image of a self-healing material based on phase separation can be seen in Figure 2.13. In this figure, the small spherical depressions represent the phase-separated domains and the large spherical hollow was a microcapsule that was filled with a polymerization initiator for the phase-separated healing agent.

2.2.6 Microcapsule Characterization

Microcapsule characterization encompasses a broad range of tasks from simple confirmation of capsule formation to quantitative analysis of mechanical performance. The most basic characterization is the 'crush-test'.

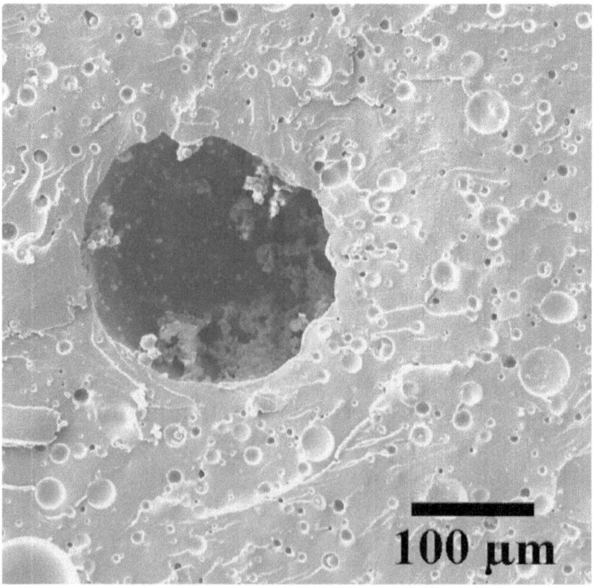

Figure 2.13 Fracture surface image showing phase separated domains and a fractured microcapsule. Reproduced from ref. 11 with permission from John Wiley & Sons, Copyright 2006 Wiley-VCH Verlag GmbH & Co. KGaA, Weinheim

Dried capsules are placed between two microscope slides and crushed to verify release. These tests coupled with optical microscopy are the primary methods by which capsule formation is ascertained. More detailed testing employs a range of techniques such as thermogravimetric analysis (TGA), differential scanning calorimetry (DSC), elemental analysis, electron microscopy, and mechanical testing which are described sequentially below.

TGA is typically used to determine what percentage of a given capsule batch is core material and as a qualitative technique for ascertaining capsule quality. To perform this test, capsules (typically *ca.* 5 mg) are placed in a small crucible on a highly sensitive microbalance contained within a temperature-controlled furnace. Depending on the oxidative stability of the capsule and core this test can be conducted in air or an inert atmosphere, an inert atmosphere is typical. As the capsules are heated, changes in mass are recorded and can be compared to control runs using pure core and shell materials. Alternatively, the TGA furnace can be coupled to a mass spectrometer to enable a more quantitative analysis of the boil or burn off products. The shape of the TGA mass loss curve is frequently used to estimate capsule quality. Ideal capsules exhibit minimal or no mass loss prior to the boiling point of the core and then a large drop as the core material vaporizes and bursts the capsules. Poor capsules are associated with TGA traces that begin to lose mass continuously as soon as the heating cycle is begun. A comparison of good and poor capsule TGA traces is shown schematically in Figure 2.14.

Differential scanning calorimetry (DSC) has been used by several groups to confirm reactivity of the encapsulated materials.[34–36] This experiment is performed by placing a small amount of crushed capsules (typically *ca.* 5 mg) within a metallic crucible. The sample crucible and an empty reference crucible are located within a furnace and analyzed using precise thermocouples. As the temperature is increased any discrepancy between the temperatures of the two thermocouples is directly related to the heat generation

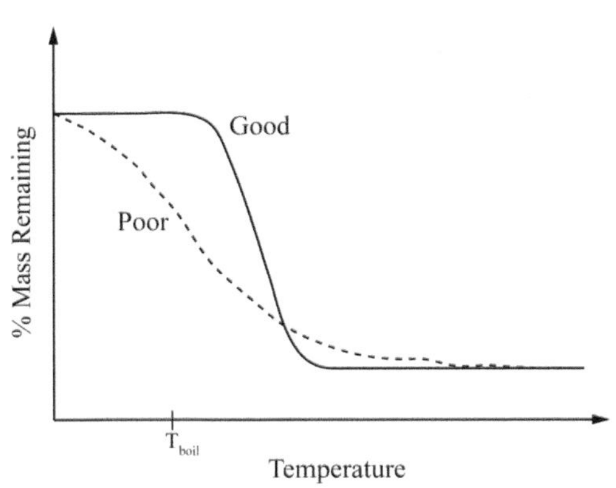

Figure 2.14 Schematic representation of a "Good" (solid line) and a "Poor" (dashed line) TGA trace of microcapsules.

caused by chemical or phase changes occurring within the sample crucible. Conceptually, this technique should be very powerful to determine the chemical activity of the encapsulated materials, however, in practice, phase transitions within the capsule shell and side reactions can mask and can obfuscate what should otherwise be clear reaction peaks.

Mechanical testing of microcapsules has also been performed to investigate capsule stability and strength. Keller and Sottos adapted a microcompression experiment to investigate capsule burst strength and to extract a modulus of the shell wall for capsules produced by *in situ* UF microencapsulation.[37] This experiment is performed by compressing a single microcapsule between rigid platens until failure, and an example load–displacement plot of the compression of a UF capsule is shown in Figure 2.15. Burst strength is determined directly from the maximum load experienced during the test and modulus is extracted by comparison with one of several available models for capsule compression. This approach has been used to investigate the mechanical properties of microcapsules formed by the UF process,[37] interfacial polymerization,[14] and by some of the more advanced processes described above.[16]

Microscopy plays a large role in the characterization of microcapsules. Optical microscopy is typically utilized for capsule size distribution analysis as well as confirmation of capsule formation. Most microcapsule shell-walls are made of amorphous polymers and are, therefore, somewhat transparent, as can be seen in Figure 2.15. This transparency enables the use of fluorescent dyes to investigate capsule release behavior. However, most microcapsules are relatively large and the limited depth of focus of optical microscopes requires that scanning electron microscopy be used for more complete characterization of capsule morphology and shell-wall thickness. Transmission electron microscopy is required for the analysis of the smallest microcapsules (*i.e.* <10 μm).

2.3 Chemistry of Encapsulated Self-healing Materials

The main body of development on encapsulation-based self-healing materials has focused on the expansion of the range of available healing chemistries. Suitable healing chemistries have the following properties:–

1. Amenable to encapsulation;
2. Stable during extended periods of inactivity;
3. Relatively fast reaction rates, *i.e.* hours not days;
4. Near room temperature reactions;
5. One liquid constituent and
6. Reaction is insensitive to air and moisture.

These six characteristics are general requirements for the selection of an appropriate healing chemistry. Note that characteristic 6 (above) may be considered a guideline, as several approaches have used environmental

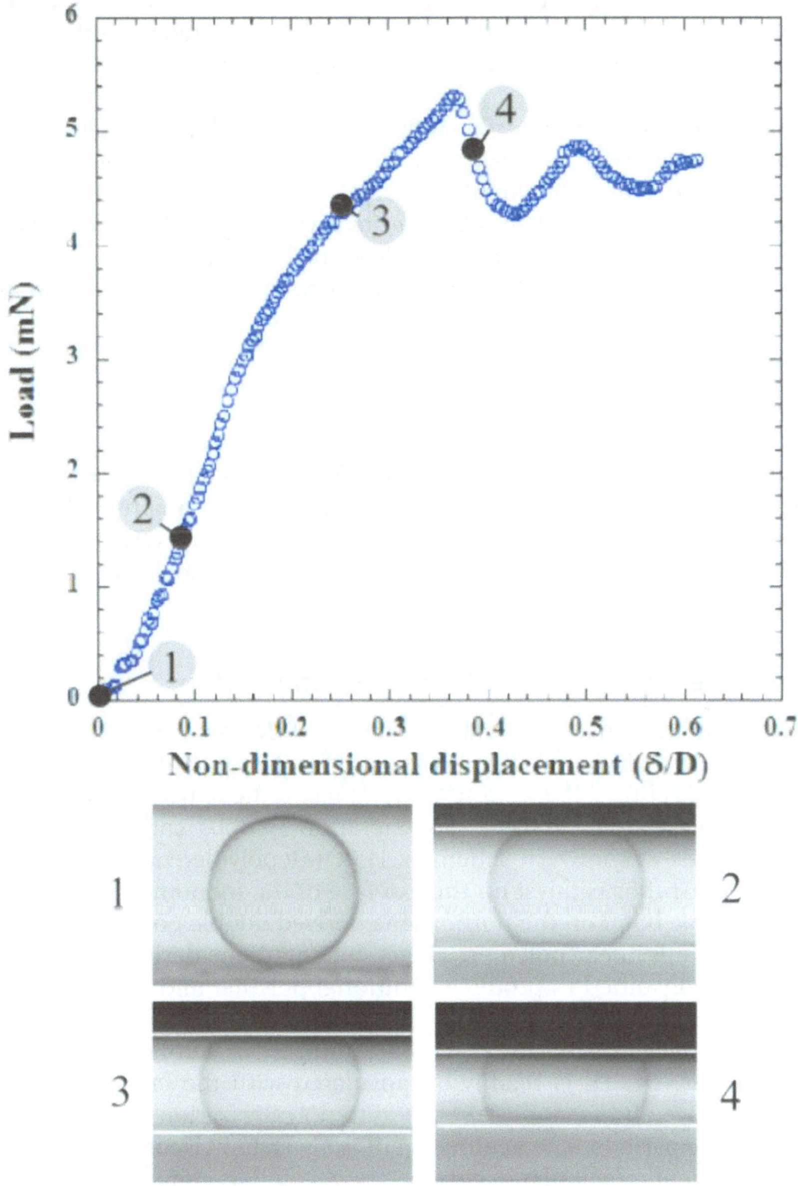

Figure 2.15 Load–displacement plot of the compression of a single microcapsule containing DCPD. The inset images were taken at the points indicated on the displacement curve, where 1 is pre-compression, 2 is during the elastic region, 3 is post 'yield' and 4 is post burst. Reproduced from ref. 37 with permission from Springer Nature, Copyright 2006.

moisture as the trigger for healing. The requirement of be able to encapsulate is typically the most stringent and has prevented several seemingly ideal healing chemistries from being adopted, such as the use of very reactive cyanoacrylates.

While the details of the performance testing and characterization of self-healing materials will be discussed later in this chapter, the concept of healing efficiency to help quantify the relative benefits and drawbacks of individual healing chemistries also merits coverage (see Chapter 1). The use of healing efficiency to quantitatively describe the effect of self-healing was first proposed by Wool and O'Conner.[1] Healing efficiency η is defined as the recovery of some defined material property:–

$$\eta = \frac{\Gamma_{\text{healed}}}{\Gamma_{\text{initial}}} \times 100\% \quad (2.1)$$

The particular material property can be fracture toughness, modulus, ultimate strength, or any other measurable material parameter. The techniques that have been developed to investigate healing efficiency will be discussed in detail in the next section and have also been outlined in Chapter 1.

2.3.1 Ring-Opening-Metathesis Polymerization

Ring-Opening-Metathesis Polymerization (ROMP) was the first healing chemistry in a fully autonomic self-healing polymer. This system is composed of a cross-linkable monomer, dicyclopentadiene (DCPD), and a solid-phase polymerization catalyst, 1st-generation Grubbs catalyst.[2] ROMP of DCPD proceeds as shown in Scheme 2.1. ROMP polymerization proceeds by attack of the Grubbs catalyst on the exo face of the monomer and opens the strained double bond on the norbornene-derived double bond. This reaction forms the initial, linear polymer. The cyclopentadiene-derived double bond will then be consumed to produce additional polymer and to crosslink with other, growing chains.

This healing chemistry has several advantages for self-healing materials. The monomer, DCPD, is readily encapsulated with the *in situ* UF process described above and is stable to both auto-polymerization and degradation over long time periods. The healing agent is also a low viscosity liquid, which enables efficient crack-plane coverage *via* capillary forces. From a cost

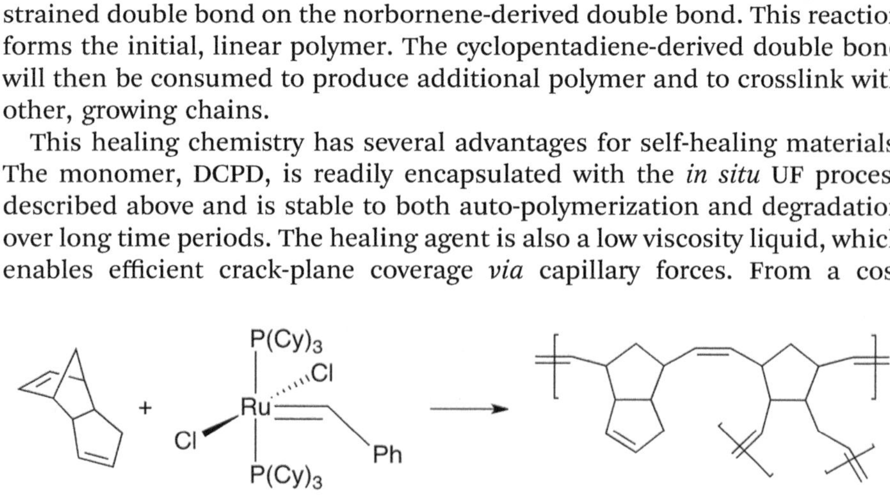

Scheme 2.1 ROMP of the cross-linkable monomer DCPD by a Grubbs catalyst.

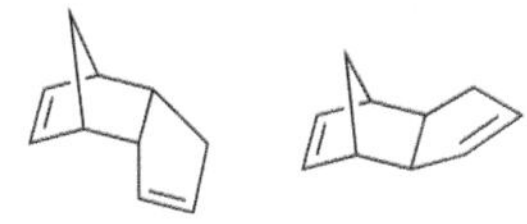

Figure 2.16 Isomers of dicyclopentadiene (a) *endo* and (b) *exo*.

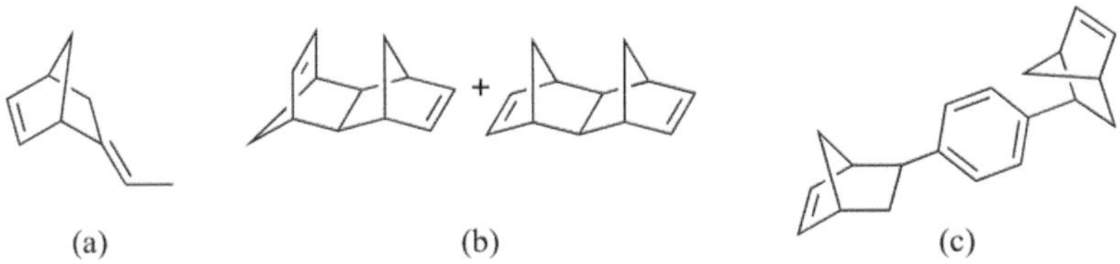

Figure 2.17 Healing agents studied as potential self-healing chemistries. Ethylene norbornene (a) was blended with the crosslinkers (b) and (c).

standpoint, DCPD is also relatively inexpensive, as fortuitously it is a by-product of the steam cracking process. Grubbs catalysts are relatively expensive, but only small amounts are required to successfully initiate and sustain ROMP.

The kinetics of the ROMP polymerization using first generation Grubbs catalyst were investigated from a mechanistic approach by Rule and Moore.[38] In this study, the relative reactivity of the *endo* and *exo* isomers of DCPD, shown in Figure 2.16, were investigated.

In this study, the *exo* isomer was shown to polymerize faster than the *endo* form. Mechanistically, this reaction rate difference was attributed to steric hindrance of an approaching monomer by the ruthenium center. The kinetics of the polymerization reaction was studied from a cure-kinetic approach by Kessler and White.[39] This study investigated the impact of catalyst concentration on cure rates and found, as expected, greater concentrations of catalyst increased the cure rate and the ultimate degree of cure.

In addition to the studies of *endo* and *exo*-DCPD, other groups have explored new ROMP-active monomer blends as candidates for healing agents. Liu and co-workers investigated the use of ethylene norbornene (ENB) blended with two norbornene-derived cross-linkers that are shown in Figure 2.17.[40]

These monomer blends were evaluated using a novel rheokinetic technique that was based on coating the surfaces of a parallel plate fixture with epoxy containing a Grubbs catalyst. Sanding the coating exposed active catalyst and then the target healing agents were applied and the curing polymer film was analyzed by rheology. This experimental set-up mimicked the conditions on the crack-plane and allowed for a simulation of the real-world conditions of polymerization for a healing agent. Using this experimental approach, the addition of cross-linkers was found to significantly

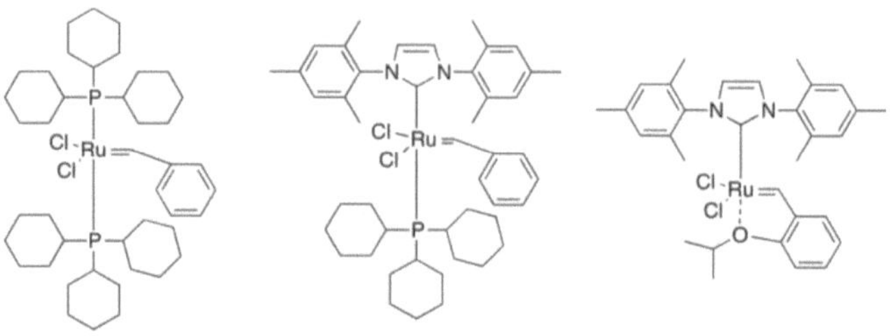

A B C

Figure 2.18 Variants of Grubbs catalysts: (A) 1st generation Grubbs catalyst; (B) 2nd generation Grubbs catalyst; (C) Hoveyda–Grubbs catalyst.

improve the cure time of the healing agents, but attempts were not then made to use these new healing agents in self-healing materials.

Several derivatives of Grubbs catalyst have been developed since the introduction of the first generation. The ROMP activity of the second-generation Grubbs catalyst and Hoveyda–Grubbs catalyst (shown in Figure 2.18) were investigated by Wilson and co-workers using DCPD and two other ROMP-active monomers, 5-ethylidene-2-norbornene (ENB) and 5-norbornene-2-carboxylic acid (NCA). These catalysts exhibit improved thermal and environmental stabilities, key parameters for the production of long-lived healing materials.[41]

In the solution-based kinetic studies, the next generation catalysts performed significantly better. However, in mechanical testing of fully *in situ* healing, the next generation systems did not perform as well as the first-generation catalyst. The primary reason for this discrepancy was found to be the competition between the dissolution of the catalyst and the polymerization of the surrounding monomer. The higher reaction rates for the second generation of catalysts led to the formation of a cross-linked shell around undissolved catalyst particles. This shell prevented catalyst from dissolving in the surrounding monomer and significantly limited the available catalyst for polymerization. As observed by Kessler and White, catalyst concentration was a critical parameter for determining the rate and ultimate cure state of the DCPD/Grubbs healing system.[39]

Based on these results, Jones and co-workers used a model system to study catalyst dissolution and polymerization kinetics.[42] First generation Grubbs catalysts with several physical morphologies were synthesized using a solvent exchange, freeze-drying, and precipitation approach. As shown in Figure 2.19(D), catalyst particles as received from the supplier (in this case Sigma-Aldrich), are large cubical crystals (note the shift in scale bar in Figure 2.19(D)). These larger crystals do provide some benefit, in that the larger particles allow for the formation of *in situ* encapsulation with a thin shell of deactivated catalyst surrounding and protecting a 'core' of active catalyst. Each of the different physical morphologies afforded different dissolution rates. The fastest dissolution rate was given by the freeze-dried morphology, Figure 2.19(C), which consisted of thin, porous wafers.

Figure 2.19 Comparison of catalyst morphologies based on various processing techniques (A) vacuum evaporation from CH_2Cl_2, (B) precipitation from CH_2Cl_2/acetone, (C) freeze-drying from benzene, and (D) as received from Sigma-Aldrich. Note the scale bar difference in image (D). Reproduced from ref. 42 with permission from the American Chemical Society, Copyright 2006.

Unfortunately, the fast-dissolving nature also implied that there was a large-surface area to volume ratio and was therefore more easily deactivated by curing agent contact.[42]

In addition to catalyst processing, Maudlin and Kessler studied the impact of the healing agent on the kinetics of catalyst dissolution in self-healing systems. By investigating a series of norbornene derivatives using a Hansen solubility parameter approach, these researchers were able to determine specific pendent functionalities that enhanced the dissolution of the Grubbs catalyst. However, the Hansen parameters were found to be non-predictive when comparing monomer blends with significantly different viscosities.[43]

In order to reduce the deactivation of the catalyst by the amine curing agent in epoxy, Rule and co-workers developed a wax-encapsulation procedure to protect the catalyst.[43] Ground catalyst was suspended in a molten paraffin wax that was then added to a heated, stirred water bath that contained a surfactant. After the wax was emulsified, a cold-water charge was added to rapidly lower the bath temperature and solidify the molten wax into catalyst-laden particles. These particles were then incorporated into a self-healing composite in the same manner as microcapsules or catalyst particles. The catalyst was activated by the healing agent dissolving the surrounding paraffin encapsulation and then dissolving the now exposed catalyst particle. Wax encapsulation protected the catalyst from the amine curing agent and allowed for an order of magnitude reduction in catalyst loading, from 2.5 wt% to 0.25 wt%, with little impact on healing efficiency. Introducing paraffin wax into the curing healing agent did change the mechanical behavior of the healed specimen, as wax is a plasticizer for poly(DCPD). The plasticization introduced non-linear fracture behavior, which will be discussed in greater detail below.

Grubbs catalyst-based ROMP systems are extremely flexible and effective in self-healing systems, but suffer from a significant drawback, the cost of the catalyst. Currently, the cost of a Grubbs catalyst is roughly twice the cost of an equivalent amount of gold. While the amount of catalyst required is relatively small, the cost can be significant for low-value, small margin composite and polymer consumer goods. In an effort to develop a cost-effective ROMP-based self-healing system, Kamphaus and co-workers investigated the use of a Tungsten-based catalyst system as a direct replacement for the more costly Grubbs system.[45] The system was based on tungsten hexachloride with a phenylacetylene co-activator and required the use of the *exo*-isomer of DCPD to achieve useful polymerization rates. Unfortunately, as a result of a combination of chemical incompatibility with epoxy resin and a tendency to clump in the matrix, healing efficiencies were considerably lower when compared to the original DCPD/Grubbs chemistry.

2.3.2 Siloxane-based Healing Systems

Siloxane-based healing chemistries are divided into two classes of polymerization systems. A polycondensation polymerization utilizes an alkoxy-end capped siloxane prepolymer and a hydroxyl-end capped cross-linker. This condensation reaction is initiated and mediated by one of several catalysts. Cho and co-workers explored this system over a course of several papers using a polycondensation approach shown in Scheme 2.2.[11,46]

In this research, the catalyst was an organotin compound, dilauryl dibutyl tin, and it was encapsulated using an interfacial polyurethane procedure. The resin and cross-linking material were either encapsulated together within a single UF microcapsule or mixed directly into the matrix material to form a phase-separated material as described above. Healing efficiencies for this system, when used to repair an epoxy matrix, were low, likely the result of the significant mismatch between the material properties of the host matrix and healing agent. Notably, however, this healing chemistry has demonstrated the capability of self-healing when immersed in water.[46]

Hydrosilylation was utilized by Keller and co-workers for the healing of an elastomeric poly(dimethylsiloxane).[47] The hydrosilylation-based healing chemistries function by the cross-linking of a vinyl-functionalized prepolymer with a hydrosilane functionalized cross-linking agent. A platinum-based

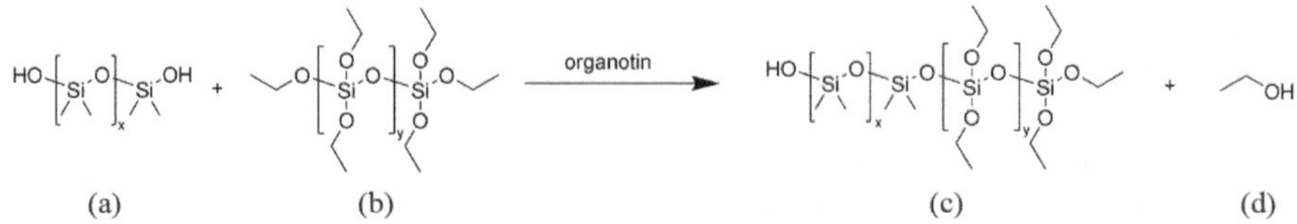

Scheme 2.2 Polycondensation of hydroxyl terminated siloxane 'a' and alkoxy-end-capped siloxane 'b' used to produce a polymer system capable of healing whilst submerged in water.

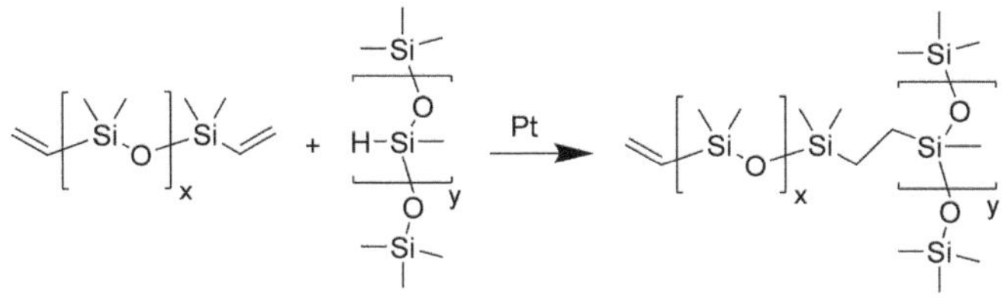

Scheme 2.3 Platinum catalyzed hydrosilation chemistry used to produce a healable elastomeric siloxane-based polymer.

catalyst, typically Kardstedt's catalyst, is present in small amounts in one of the two liquid components of the healing system.

Scheme 2.3 shows the cross-linking reaction for this chemistry, which has two unique characteristics:

1. Healing was demonstrated using a chemistry that required the mixing of two liquid components on the crack plane.
2. The healing chemistry also produced the same polymer as the matrix polymer.

The matching of the healing agent-derived polymer and the matrix material led to extremely high healing efficiencies.[47] Another feature of this healing system is that the released healing agent was able to take advantage of unreacted monomer, or latent-functionality, present in the matrix to accomplish healing. Several healing systems have since attempted to utilize this same approach to improve healing efficiencies.[48]

2.3.3 Radical-initiated Healing Chemistries

Wilson and co-workers were the first to investigate the potential of acrylate systems as healing chemistries by studying the kinetics of the various peroxide-based polymerization initiators.[49] In this initial study, five radical polymerization initiators were investigated as shown in Figure 2.20.

Of the peroxide initiators that were studied, two were amenable to encapsulation within a urea-formaldehyde shell, benzoyl peroxide (BPO) and lauroyl peroxide (LPO). Microencapsulation was performed by dissolving the initiators in a solvent – phenyl acetate for BPO and hexyl acetate for LPO and then encapsulated as described above. Healing was investigated by control experiments but an *in situ* system was not tested.

The active chain ends of polymers formed by living radical polymerizations, such as atom transfer radical polymerization (ATRP) and reversible addition-fragmentation chain transfer polymerization (RAFT), have also been exploited as initiators for radical-based self-healing chemistries. These healing chemistries take advantage of the long half-lives of end-chain radicals for initiating polymerization of a released healing agent. Two related

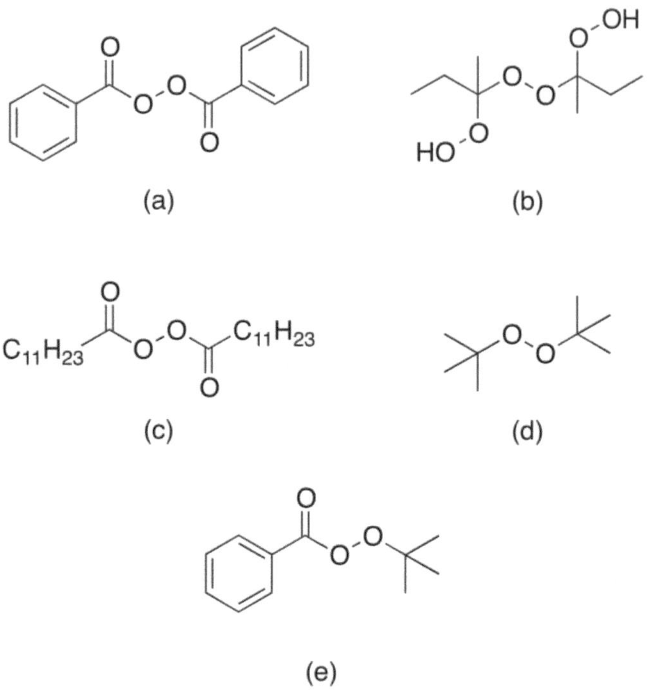

Figure 2.20 Peroxide initiators studied for acrylate-based self-healing (a) benzoyl peroxide, (b) methyl ethyl ketone peroxide, (c) lauroyl peroxide, (d) *tert*-butyl peroxybenzoate and (e) tert-butyl peroxide.

papers have been published that utilize the living radicals that are found within a polymer produced by ATRP or RAFT polymerization.[50,51] In the first report of this technique, a poly(methyl methacrylate) polymer was formed using an ATRP polymerization and an acrylate healing agent, glycidyl methacrylate (GMA), was encapsulated and dispersed in the monomer solution prior to polymerization. After the capsules were ruptured by a crack, the GMA polymerization was initiated by the 'living radicals' within the polymer.[50] Impressive healing efficiencies approaching 100% were reported. However, in this system, healing had to be carried out under an inert atmosphere, as exposure to air quickly deactivated near-surface radicals. Attempts to minimize this drawback include investigating the use of a bulk matrix produced by RAFT polymerization, in which the living radicals are more stable in air.[51] This healing system utilized the same microencapsulated healing agent GMA and also exhibited high healing efficiencies.

2.3.4 Isocyanate-based Healing Systems

Isocyanate systems would seem to be ideal for self-healing applications as they exhibit fast reaction rates frequently without any initiator or catalyst utilizing environmental moisture as a cross-linker. Unfortunately, the very characteristics that make this system ideal are also the same that typically prevent isocyanates from being encapsulated. Key reports in the literature have described how this limitation can be successfully overcome to produce microcapsules containing highly reactive isocyanates. The first demonstration of liquid isocyanate encapsulation utilized an aliphatic isocyanate, isophorone diisocyanate (IPDI), encapsulated using a prepolymer based upon toluene diisocyanate (TDI), an aromatic diisocyanate. By taking advantage of the large disparity in reaction rates between the aromatic and aliphatic diisocyanates, the liquid IPDI was encapsulated successfully within a polyurethane shell using interfacial polymerization.[14] However, in this case, the reactivity of IPDI released from the microcapsule was demonstrated, but did not facilitate self-healing. Subsequently, Huang and Yang successfully adapted this encapsulation approach to produce microcapsules containing hexamethylene diisocyanate by using a methyldiphenyl diisocyanate prepolymer for the interfacial polymerization.[15] This system did demonstrate the qualitative capability to self-heal.

2.3.5 Epoxy-based Healing Systems

Epoxy-based healing systems were the first self-healing chemistries that were based on the incorporation of a secondary phase in a host material. Zako and Takano incorporated solid particles of a high-temperature curing epoxy material within a room temperature curing epoxy composite.[52] After damage, heat, between 110 and 140 °C, was applied to melt the embedded particles, which then flowed into the crack-plane and cured. However, the concept of healing efficiency had not gained widespread understanding at that time and so the results were presented as a simple comparison of healing and non-healing performance.

Since this first demonstration, several groups have reported healing based on the ring-opening polymerization of epoxy systems. A significant proportion of the research investigating the performance of hollow-fiber-based self-healing materials has been performed using epoxy resin and amine curing agent-based healing systems.[20,53–58] These systems avoid the complexity and difficulty of encapsulating amine curing agents, as the hollow-fiber encapsulation procedure is physical in nature. Fibers are embedded in a material and then the healing agents are infiltrated using a vacuum process. Most hollow-fiber systems directly encapsulate commercial room-temperature curing epoxies as healing agents. Healing efficiencies have been assessed by low-speed impact and can be very high, up to 91% recovery of flexural strength.[57]

The first report of successful microcapsule-based healing using epoxy chemistry was based on a latent hardener complex $CuBr_2(2\text{-MeIM})_4$.[59] To accomplish healing, an epoxy prepolymer was encapsulated within urea-formaldehyde microcapsules and was then dispersed in an epoxy resin. The epoxy matrix was blended with the latent hardener complex prior to curing.

This system required the application of heat, 130 °C, to initiate and complete the polymerization of the released epoxy. Healing efficiencies were between 40% and 90% depending on concentrations of the latent hardener and the microcapsules. Other catalytic curing agents have been encapsulated and utilized as initiators for the polymerization of epoxy prepolymers. A microcapsule-based material was recently demonstrated using scandium triflate as the catalytic curing agent.[60] A mercaptan healing agent has also been encapsulated and used in self-healing polymers and composites and demonstrated healing at room temperature.[61]

Recently amine-based hardeners have been successfully microencapsulated using a particle-stabilized reverse water-in-oil emulsion.[13] The encapsulation of amines has proven to be difficult as most useful amine-based hardeners are readily water-soluble and so not directly amenable to the encapsulation approaches described above. Interfacial encapsulation was used to encapsulate an amine adduct and the emulsion was stabilized by using nanoparticles. Healing was demonstrated in a self-healing adhesive formulation assessed by fracture and fatigue.[62]

2.3.6 "Click-chemistry"-based Healing Agents

Click chemistry is a philosophy rather than a specific chemical system, which requires that a click reaction has the features of high reactivity, specificity, stable products, and few or innocuous by-products.[63] These reactions are exemplified by a few reactions, most notably the azide-alkyne Huisgen (3 + 2) cycloaddition process that is shown in Scheme 2.4.

While no direct reports of successful self-healing utilizing this chemistry are available in the literature, there has been successful encapsulation of a click-chemistry system with alkyne and azide healing agents as shown in Figure 2.21.[64]

These monomers were encapsulated in a UF shell using the same approach as the one outlined by Brown and co-workers.[5] The reaction between the monomers shown in Figure 2.21 was catalyzed by a copper compound, $CuBr(PPh_3)_3$, that was embedded in the host matrix. To demonstrate reactivity, the microcapsules were mixed into a lightly cross-linked poly(isobutylene) matrix and deformed to rupture the capsules. Reactivity was assessed by measuring the increase in modulus that was generated by the cross-linking of the rubbery polymer with the released capsule contents.

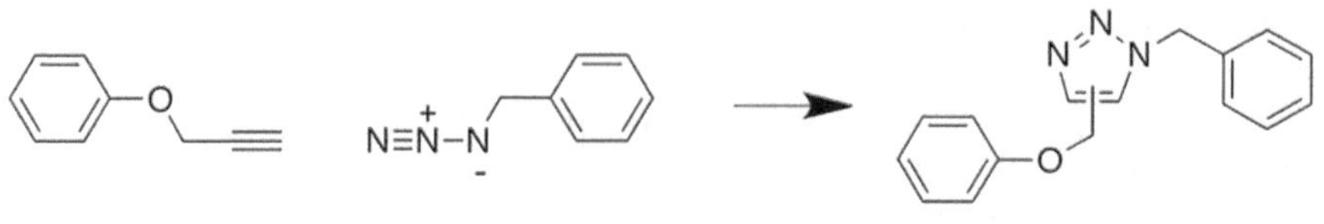

Scheme 2.4 Example of the Huisgen (3 + 2) cycloaddition reaction, frequently now referred to as a 'click' reaction.

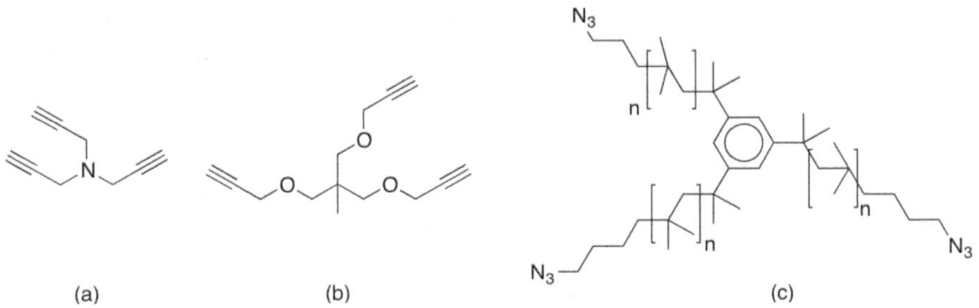

Figure 2.21 Healing agents for the proposed azide/alkyne "click-chemistry"-based self-healing material.

2.3.7 Solvent or Plasticizer-based Healing Chemistries

Solvent-based healing chemistries provide an interesting alternative to the heterogeneous adhesion approaches described above and the first studies of crack and damage healing were performed using solvents.[1] However, these early studies required the use of higher temperatures or pressures and long time-spans to fully recover material properties. Unlike the chemistries described above that involve adhesive bonding of a polymerized healing agent to a host polymer, solvent-based chemistries rely on promoting the local interdiffusion of the host polymer across the facture surface. Several reports of successful self-healing have been published whereby a solvent,[36,65–68] or combination of solvent and reactive agent,[48] were used as the healing agent. Solvents that have been successful in epoxy-based matrices include chlorobenzene, phenyl acetate, and ethylphenyl acetate.[23,48] Healing efficiencies are typically high in these approaches and can exceed 90%.

In addition to the purely solvent-promoted healing, two hybrid solvent/latent functionality approaches have been reported. By mixing a small amount of epoxy resin material into a solvent and then encapsulating the resulting solution, healing efficiencies of 100% were reported.[48] The proposed mechanism of healing in this system was a combination of solvent promoted chain interdiffusion and the harvesting of latent amine functionality in the matrix. This proposed mechanism was confirmed by heat-treating the self-healing material to ensure complete consumption of free amine within the cured matrix. After the heat treatment, the healing efficiency dropped significantly. As a further development of this approach, Meng *et al.* used encapsulated glycidyl methacrylate to take advantage of a combination of non-covalent interaction and covalent bonding with free amine within the host matrix.[36] This system also demonstrated 100% healing efficiencies.

2.3.8 Summary of Encapsulation based Self-healing Materials

Since the initial introduction of ROMP-based healing chemistry, researchers have mined the polymer science literature for environmentally stable chemistries that enable successful encapsulation and healing chemistries. Many well-established chemistries have been implemented, but to date a 'universal' encapsulation system that offers high healing efficiencies for all bulk matrix materials has yet to be achieved. Further developments include the potential of using microencapsulated nano-particles as an active healing chemistry,[68] however, the perfect combination of stability, reaction rate, and cost has not been achieved at the moment. Each of the chemistries described above is useful in specific circumstances and both the operating regime and matrix material will dictate selection when formulating a self-healing polymer or composite.

2.4 Performance and Testing of Microcapsule-based Self-healing Materials

As exemplified above, the performance testing of self-healing materials is typically based on the measurement of the recovery of some critical material parameter, *i.e.* the healing efficiency (see Chapter 1). In many of the encapsulation-based healing polymer systems the fracture toughness has been the most frequently measured material property.[2,69–71] However, modulus,[52] impact strength,[72,73] and tear strength[47] have also been utilized to assess healing efficiencies. In this section the mechanical testing and performance of the various self-healing systems and encapsulation-based materials that have been developed will be discussed. For ease of navigation, this section will be structured around the types of testing performed, such as fracture, impact, or tear.

As with all materials testing studies, appropriate and accurate control experiments are critical for the characterization of self-healing materials. This is especially true in self-healing materials, as the mechanical testing does just not represent a physical assessment, but also a mechanism investigation (see Chapter 1). Most self-healing materials testing of multi-component healing systems will have a test regimen for each of the individual components as well as the basic matrix material to investigate the contribution of each phase and to eliminate any potential contribution of spontaneous self-healing of the matrix. This last point is particularly critical for elastomers or other polymers with high values of surface adhesion and relatively large free volume. In addition to the individual components, two specialized control tests, the injected heal and the self-activated heal, are usually performed. The injected heal uses premixed healing agent to investigate the behavior of the healing chemistry when mixed in ideal stoichiometric quantities The self-activated heal applies to multi-component systems and has the polymerization catalyst or initiator present in the host polymer and an unmixed healing agent which is then injected onto the damage plane.

2.4.1 Fracture Testing

The earliest research in self-healing utilized a fracture-based material characterization approach. Recovery of fracture toughness was assessed using a double-cantilevered beam (DCB) geometry for a lightly cross-linked poly(butadiene) matrix.[1] While the standard DCB specimen is well-characterized and understood from a fracture-mechanics standpoint, this geometry has one major drawback, the need to know the location of the crack-tip in order to calculate the fracture toughness. Determining the location of the effective crack-tip in the initial test is a straightforward process, but during the healed test finding the crack-tip can become difficult, if not impossible. The solution to this practical issue arose from the work of Mostovoey, Crosley, and Ripling who, in 1967, realized that the dependence of fracture energy on specimen compliance could be exploited to fabricate specimens that are crack length independent.[74] The relationship between specimen compliance $C(a)$, which is dependent on crack-length, and fracture energy G is given by eqn (2.2):

$$G = \alpha \frac{\mathrm{d}C(a)}{\mathrm{d}a} \tag{2.2}$$

where a is the crack length and α is a parameter that depends on specimen width and failure load. For linear elastic materials, a valid assumption for many engineering polymers, the functional form of C can be determined using beam theory. In the case of the typical double cantilevered beam specimen (DCB):[75]

$$G = \frac{12P^2a^3}{Eh^3B} \tag{2.3}$$

where P is the applied load, a is the crack length, E is the elastic modulus of the material, h is the height of the specimen and B is the width of the specimen. As can be seen from eqn (2.3), choosing the appropriate height profile h, or width profile B, the fracture energy, and therefore fracture toughness, can be made constant with respect to crack length a. Fracture toughness is related to fracture energy by the relationship given in eqn (2.4).

$$G = \begin{cases} \dfrac{1-\nu^2}{E}K^2 : \text{Plane Strain} \\ \dfrac{K^2}{E} : \text{Plane Stress} \end{cases} \tag{2.4}$$

Specimens that are constructed by keeping the width constant and varying the height profile are referred to as height-tapered double cantilevered beam (TDCB) specimens. When the height is constant and the width is tapered, these are referred to as width-tapered double cantilevered beam (WTDCB) specimens. The use of crack-length independent specimens represented a critical step forward in the unambiguous characterization of the recovery of fracture properties. For the initial studies of the DCPD/Grubbs systems,[71] the specimen was a height tapered fracture specimen, as shown in Figure 2.22.

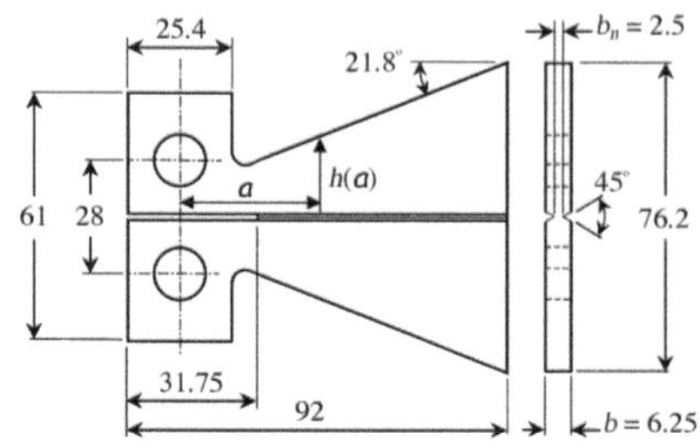

Figure 2.22 Height-tapered double cantilevered beam specimen (TDCB); all dimensions are in mm. Reproduced from ref. 71 with permission from Springer Nature, Copyright 2002.

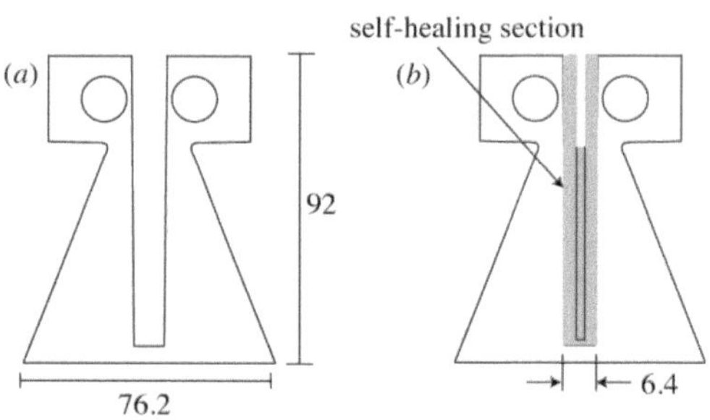

Figure 2.23 Localized self-healing TDCB, all dimensions are in mm.

2.4.1.1 *Quasi-static Loading*

Using the specimen with the geometry specified in Figure 2.21, healing efficiencies greater than 90% were obtained for an optimized system using a DCPD/Grubbs catalyst.[58] Since this initial report, the TDCB specimen has been employed to characterize a variety of other healing systems. The GMA-based solvent/nucleophilic addition system described above was also capable of recovering over 90% of the initial fracture toughness.[36] Siloxane-based chemistries utilizing phase separation have also been investigated using TDCB specimens. The fracture-strength recovery was low as a result of the mechanical strength mismatch between the healing agent-derived polymer and the epoxy material.[11] A mercaptan/epoxy system was also investigated using TDCB specimens, but exhibited variable healing efficiencies varying from 50% to full recovery.[76] Solvent systems have also been characterized using this specimen geometry and have demonstrated healing efficiencies from 60% to 100% dependent on the particular solvent that was used.[23,48]

The investigation of the tungsten-catalyzed ROMP was carried out using a localized TDCB specimen, which allowed for a reduction in the amount of catalyst required for characterization of the healing response. A localized TDCB, shown in Figure 2.23, contains only a central section of the specimen that is self-healing, reducing the required amount of material and the overall cost of the specimen. Recent work using a catalytic triflate-based polymerization initiator for a microencapsulated epoxy resin has also been studied with a localized specimen.[60]

One of the critical aspects of any self-healing system is that the inclusion of a self-healing functionality should not negatively impact the base performance of the material. If the basic performance of the material is significantly degraded to gain the additional functionality, then a pyrrhic victory has been won. In the case of microcapsule-based composites, the addition of capsules typically improves the fracture toughness of the matrix.[77] This effect is dependent both on the diameter of the microcapsules as well as the capsule concentration as shown in Figure 2.24, for an epoxy resin containing DCPD-filled UF microcapsules. Improvements in the fracture toughness are the result of the introduction of additional energy dissipation mechanisms, which manifest on the fracture surface as tails emanating from the trailing edges of microcapsules.

Fracture testing has also been utilized to investigate the recovery of interlaminar fracture toughness in woven composite materials.[69,70] Instead of a height-tapered specimen, a width-tapered test was used to assess the recovery of fracture strength. Healing was demonstrated for the composite, but clumping of the catalyst and microcapsules hampered the maximum healing efficiencies. Elevating the temperature during cure increased the healing efficiencies, but was no longer fully autonomic. Capsules were generally relegated to the resin-rich zones between warp and weft tows of the woven

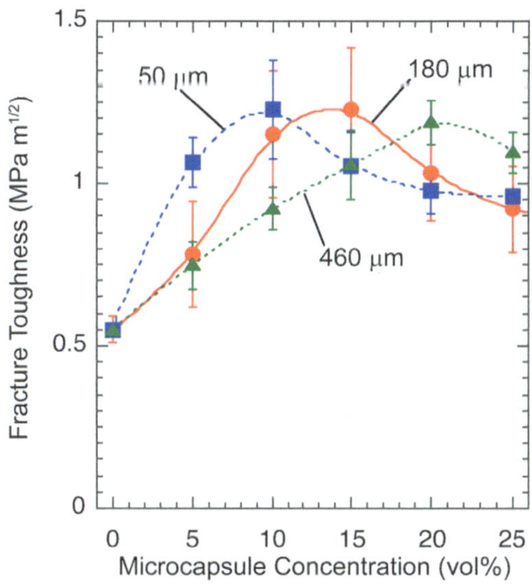

Figure 2.24 Effect of capsule size on virgin fracture toughness. Reproduced from ref. 77 with permission from Spinger Nature, Copyright 2004.

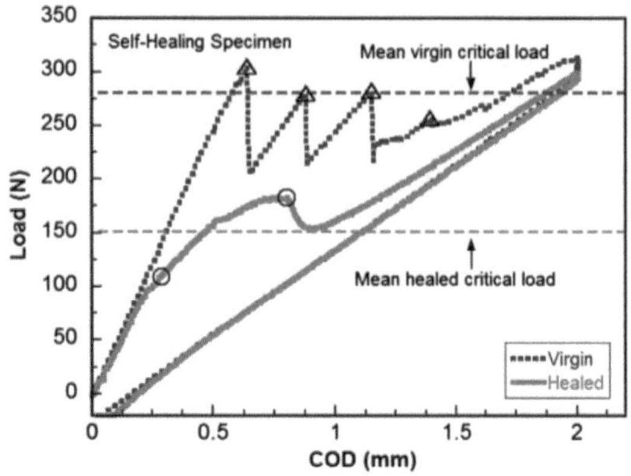

Figure 2.25 COD/load behavior of an initial and healed interfacial fracture specimen for assessing adhesive bond strength. Reproduced from ref. 62 with permission from Elsevier, Copyright 2011.

fiber reinforcement and the uneven distribution also impacted healing efficiencies.

The healing of adhesive joints has also been investigated using a fracture approach.[62] This study used the DCPD/Grubbs healing system and an epoxy-based adhesive bonding two carbon-steel adherends. Test specimens were also width-tapered as in the composite fracture testing described above. This self-healing adhesive demonstrated a healing efficiency of 56%, and an example crack-opening-displacement (COD) *versus* load plot for a self-healing adhesive system is shown in Figure 2.25.

Li *et al.* utilized a thermoplastic particle approach that required both heat and pressure to realize healing in an adhesive joint.[78] Complete recovery of the adhesive joint fracture strength was demonstrated, but the requirement of external intervention meant that the system was not truly self-healing in the manner defined in this chapter.

As described above, several healing chemistries exhibit non-linear fracture behavior, as shown in Figure 2.26. Non-linear fracture behavior can be the result of either a dissolved plasticizer, such as the wax surrounding a catalyst particle, or as the result of incomplete polymerization of the healing agent.

To test for this non-linear response, a fracture energy approach is typically adopted.[44] This method allows the calculation of healing efficiency by comparison of the areas under each load displacement curve. Areas are found by numeric integration of the data and then the healing efficiency is given by eqn (2.5):

$$\eta_{G} = \frac{\frac{U_{\text{healed}}}{b_n}\left(W - a_0^{\text{healed}}\right)}{\frac{U_{\text{intial}}}{b_n}\left(W - a_0^{\text{initial}}\right)} \tag{2.5}$$

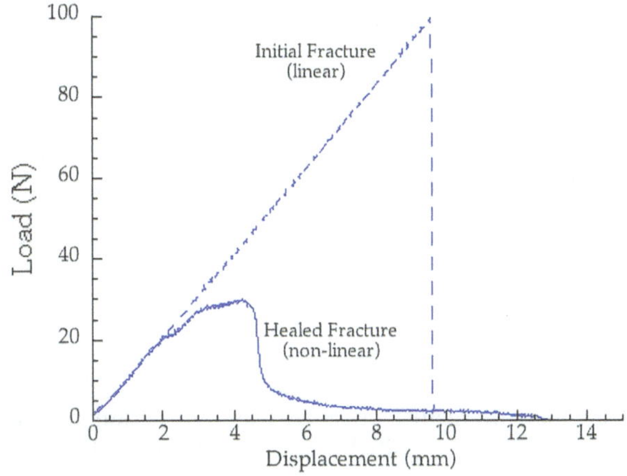

Figure 2.26 Comparison of a linear virgin fracture and a non-linear healed fracture. Image courtesy of Dr Joseph D. Rule.

where U is the strain energy, given by the area under the curve, b_n is the width of the specimen, W is the distance from the loading line to the end of the specimen, and a_0 is the initial crack length. Typically, the initial crack length is assumed to be approximately constant for the healed and unhealed specimens and the parameters W and b_n are constant. Therefore, comparison of the relative strain energies for the healed and unhealed specimens determines the healing efficiency.[44]

The diameters of the incorporated microcapsules are directly related to the size of the damage that can be healed. Rule and co-workers undertook a focused study to determine the relationship between capsule size and crack volume that could be healed.[66] A simple relationship between capsule diameter d_c, capsule mass fraction ϕ, specimen density ρ_s, and the mass of healing agent delivered m (eqn (2.6):

$$m = \rho_s \phi d_c \tag{2.6}$$

This relationship puts an upper limit on the size of damage that a given capsule diameter can address. As shown in Figure 2.27, the peak load of fracture for a healed specimen, which can be considered as a stand in for healing efficiency, is highly dependent on the volume of delivered healing agent. Thus, the capsule diameter must be tailored to address the size of the expected damage. As may be expected, smaller capsules can be effective for small microcracking, but larger capsules are required to effectively repair large damage.

In an attempt to address the size limitation of the microcapsule approach, Kirkby *et al.* introduced shape memory alloy wires into the TDCB geometry in order to provide an active process to close cracks. These wires were embedded perpendicular to the fracture plane as shown in Figure 2.28.

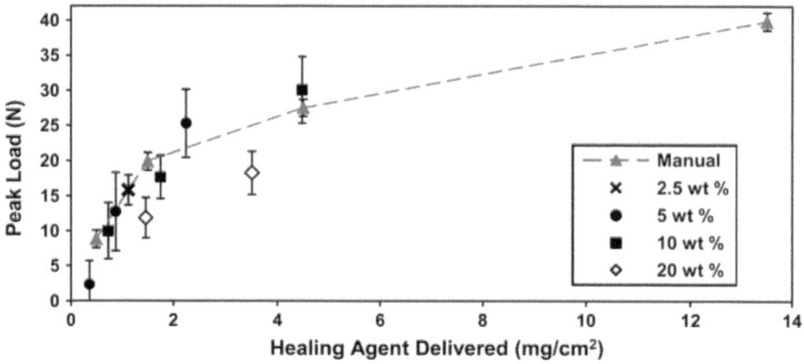

Figure 2.27 Peak load of healed specimen as a function of mass of delivered healing agent. Reproduced from ref. 79 with permission from Elsevier, Copyright 2007.

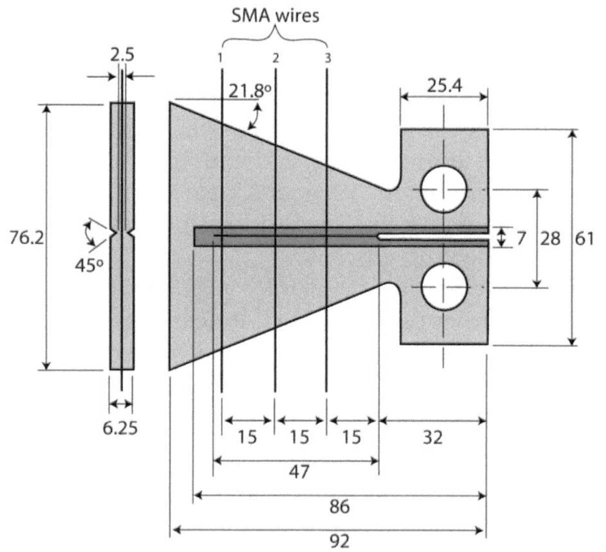

Figure 2.28 TDCB with active SMA crack closure. Reproduced from ref. 80 with permission from John Wiley & Sons, Copyright 2008 Wiley-VCH Verlag GmbH & Co. KGaA, Weinheim.

Introduction of this active closure mechanism improved average healing efficiency by a factor of approximately 1.6.[80]

2.4.1.2 *Fatigue Loading*

While quasi-static fracture recovery has been demonstrated for a variety of matrix materials, including epoxy[48,60,67,71,81] and epoxy vinyl esters,[82] self-healing has also been demonstrated in a fatigue-loading environment.

Healing efficiency for this loading is defined as the difference between the cycles to failure in the healing specimen N_{healed} when compared to the cycles to failure in a control specimen N_{control} (eqn (2.7)):

$$\eta = \frac{N_{\text{healed}} - N_{\text{control}}}{N_{\text{control}}} \tag{2.7}$$

One of the most critical differences between the behavior of self-healing materials in quasi-static fracture testing and in fatigue is the basic mechanism of healing. For self-healing of quasi-static fracture a single, adhesion-based mechanism is responsible for almost all of the recovery of fracture toughness. However, this is not the case for self-healing materials under fatigue loading, where several mechanisms operate to slow or arrest growing fatigue cracks.[83] The impact of each of the physical mechanisms initiated by the self-healing functionality on the effective stress-intensity factor at the crack-tip is shown in Figure 2.29. Prior to the rupture of any healing agent-filled microcapsules, region a, the crack tip experiences the full, applied crack-tip stress intensity. As the crack propagates and ruptures capsules, region b, liquid healing-agent is released onto the crack plane and a hydrodynamic shielding mechanism is introduced. The mere presence of a fluid on the crack surfaces can significantly reduce the apparent crack-tip stress intensity factor and lead to life extension when compared to a non-fluid filled crack tip.[84,85] As the healing-agent polymerizes, region c, the full effect of the adhesive-based healing mechanism is realized. If the adhesive bond is strong enough, or the applied loading low enough, the crack may arrest and stop propagating at this stage. However, if the loading is large enough, or the adhesive bond weak, the crack will begin to propagate again, region d, and the wedge of polymerized healing agent will introduce a

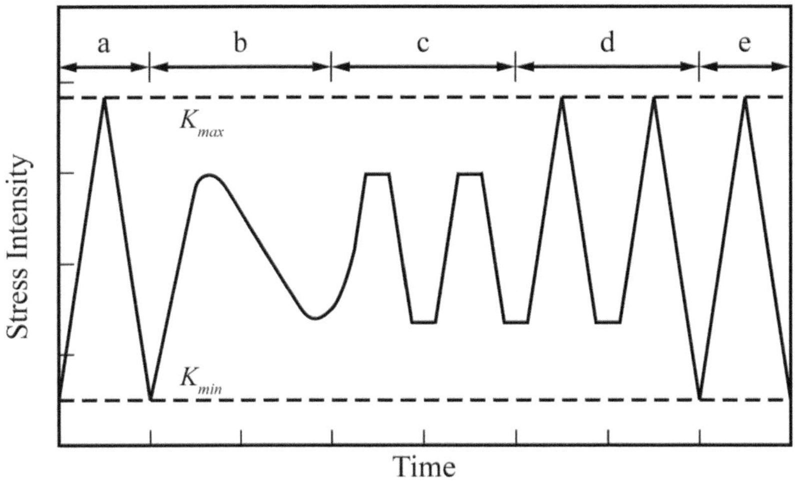

Figure 2.29 Load evolution in a self-healing fatigue test.

crack-closure mechanism. This closure will also act to slow the growth of fatigue cracks. Eventually, the crack may propagate beyond the healed zone into virgin material, region e, and the cycle will begin again.

Brown, White, and Sottos first investigated the behavior of the DCPD/Grubbs system under tensile fatigue loading in a two-part study. Initially, only the behavior of control experiments was studied. Using a precatalyzed healing agent, this first paper elucidated the evolution of the mechanisms governing fatigue crack growth rate and arrest that were outlined in Figure 2.29. Subsequently, the performance of the fully *in situ* healing system was studied.[86] The fatigue performance of the material was highly dependent on loading conditions, however, the results did demonstrate that with appropriate selection of loading, a fatigue crack could be slowed or even arrested as a result of the action of the self-healing functionality.

One of the critical findings during the investigation of fatigue behavior by Jones and co-workers, was the realization of the importance of the interaction between the mechanical kinetics of the damage, the crack, and the chemical kinetics of the healing, the polymerization.[87]

To investigate this interplay, Jones and co-workers studied the behavior of the self-healing composite under high-cycle fatigue at lowered loads. In this loading regime, a self-healing material performed as if the polymer matrix possessed a fatigue threshold, as shown in Figure 2.30(a). For a higher-load, low-cycle fatigue loading, altering the chemical kinetics of the Grubbs/DCPD system that was being tested produced significant changes in performance, as shown in Figure 2.30(b). The change in chemical kinetics was achieved by varying the concentration and physical morphology of the Grubbs catalyst present in the matrix material.[87] In Figure 2.30(b), curve 1 is a control

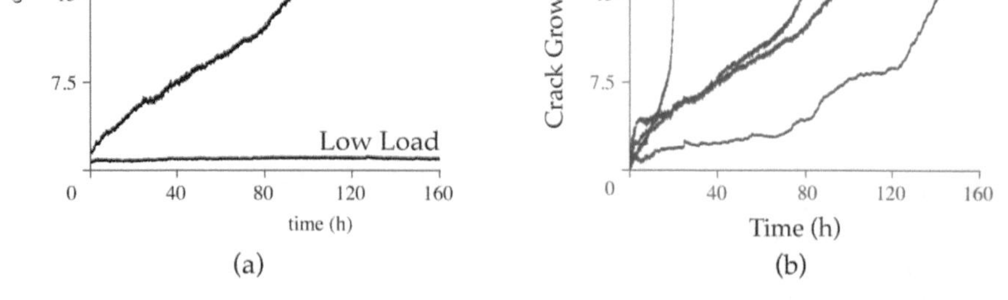

Figure 2.30 The impact of mechanical kinetics (a) and chemical kinetics (b) on the performance of a self-healing polymer under fatigue loading. Reproduced from ref. 87 with permission from the Royal Society, Copyright 2006.

specimen without catalyst, curve 2 contained 5 pph wax encapsulated Grubbs catalyst with an as-received morphology, curve 3 contained 5 pph wax encapsulated Grubbs catalyst that had been recrystallized, and curve 4 contained 5 pph wax encapsulated Grubbs catalyst that had been freeze-dried. The polymerization kinetics are influenced by the dissolution rate of the catalyst[29] and the use of the freeze-dried catalyst, which has fast dissolution kinetics, increased the polymerization rate of the released healing agent. As shown in Figure 2.30, increasing the rate of the chemical kinetics had a strong influence on the final fatigue life.

In addition to the variation of the chemical kinetics, Jones expanded on the investigation of the use of 'rest periods' initiated by Brown for improving the performance of the self-healing material.[87] These rest periods are periods of the test where the cyclic loading is paused and the chemical kinetics of polymerization are allowed to 'catch-up' with the mechanical kinetics of damage. A comparison between a self-healing composite fatigued without rest periods and a composite with equally spaced rest periods is shown in Figure 2.31. The arrows in this figure represent points in the test where rest periods were introduced. The inclusion of rest periods has a huge impact on the overall lifespan of the material and reinforces the importance of the interplay between chemical and mechanical kinetics in self-healing materials.

The studies described above were focused on the performance of a fatigue crack propagating in a bulk polymer. Fatigue has also been investigated in

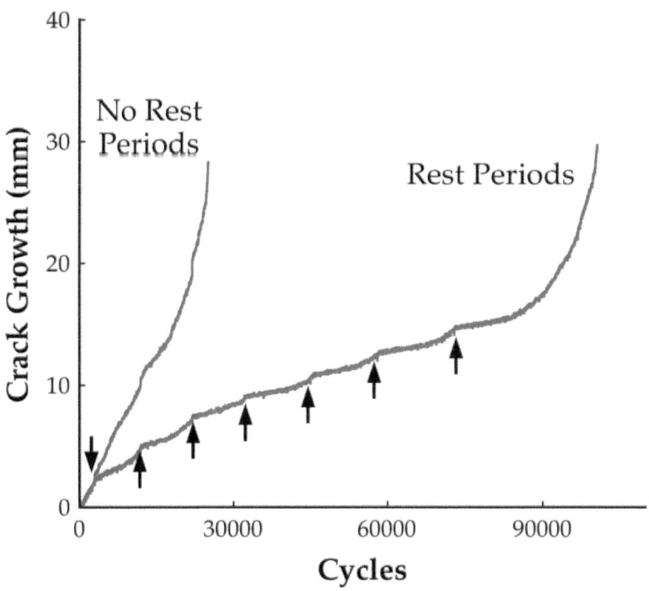

Figure 2.31 Impact of rest periods on the performance of a self-healing composite under fatigue loading. Arrows indicate rest periods. Reproduced from ref. 87 with permission from the Royal Society, Copyright 2006.

the performance of self-healing adhesive joints. Testing for these materials was performed using a similar width-tapered geometry as in the quasi-static fracture described above. Under a stress intensity ratio of $R = 0.1$, the self-healing specimen demonstrated complete arrest of the propagating fatigue crack.

The torsion fatigue of elastomers has also been investigated using the self-healing elastomer develop by Keller, White, and Sottos.[88] This mode III torsional fatigue test was chosen because of the combination of a closed crack-face that would reduce ejection of healing agent and acceptance as a model of fatigue loading for common elastomeric structures, such as the belt-edge region of tyres.[88]

The torsional stiffness curves shown in Figure 2.32 represent three control specimens and one fully *in situ* self-healing material. R-I is a non-healing control containing no microcapsules, R-II contained only resin capsules, R-III contained only cross-linker capsules, and R-IV is the fully *in situ* healing system. Both the *in situ* and cross-linker only specimens exhibited a stiffness increase at approximately the time required for the healing chemistry to gel, approximately 10 hours. The measured increase in stiffness at 10 hours represents the action of the polymerized healing agents, which slowed crack propagation. Post-testing measurements determined that the crack lengths in the two active samples were up to 24% shorter when compared to the non-healing control. This decrease in crack length was attributed to an artificial crack-closure process like that found in the tensile fatigue loading, as represented by region d in Figure 2.29.

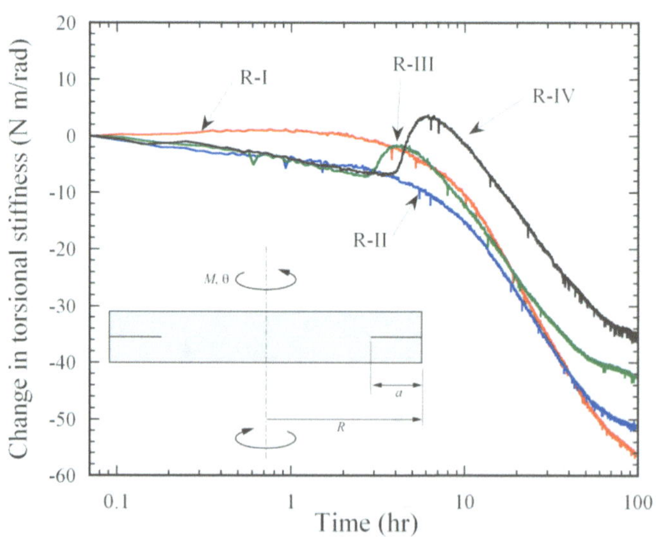

Figure 2.32 Stiffness evolution of the torsion fatigue of a self-healing PDMS elastomer. R-I is a non-healing control. R-II contained only resin capsules, R-III contained only crosslinker capsules, and R-IV is the fully *in situ* healing system. Adapted from ref. 88 with permission from Elsevier, Copyright 2008.

2.4.2 Impact/Ballistic Testing

There are two types of ballistic or impact testing that are performed on self-healing materials. Direct healing of impact damage involves a process where a self-healing material is formed into a Charpy or Izod impact specimen and then tested to failure. The second approach is to measure the recovery of a residual property, such as flexural strength, in a self-healing composite that has been damaged by impact or indentation (see Chapter 1 for further information on these techniques).

Several research groups have characterized the self-healing of damage generated by impact on composite panels using the compression-after-impact (CAI) approach. In this test a composite panel is damaged using either low-speed impact (indentation) or a drop weight tower. The panel is then loaded in a specially designed fixture and compressed until panel failure in order to assess residual strength. Of the encapsulation systems that have been studied using this approach, the hollow fiber-based materials exhibit superior healing efficiencies when compared to the microcapsule-based systems. Bond and co-workers investigated the performance of hollow-fiber-based graphite/epoxy composites.[57] These systems demonstrated excellent healing efficiencies recovering between 70% and 90% of the undamaged CAI strength depending on the force of the indentation. Microcapsule-based systems demonstrated significant recovery of compressive strength, up to 96% recovery, at low impact energies (*ca.* 18 J).[89] As the impact energy increased the healing efficiency decreased. This effect was attributed to a combination of increasing crack volume requiring more healing agent than could be delivered by the embedded microcapsules. The effect of damage size on healing performance was described in Section 2.4.2.

Izod impact testing has been used extensively to characterize the healing of the living radical promoted materials.[50,51,90] In these studies, the Izod sample, a thin rectangle clamped in cantilevered beam geometry, is impacted and completely severed. The two halves of the sample are brought back into contact and allowed a healing period under a temperature-controlled environment. In the case of the radical systems, the healing efficiencies were very high but required healing in a desiccator under argon.[90] A related test, Charpy impact, was used to investigate the healing of an epoxy resin blended with a thermoplastic,[91] which also saw high healing efficiencies, but elevated temperatures and pressure were required.

2.4.3 Tear Testing

Healing of elastomeric materials is frequently assessed by a tear-specimen. Tear specimens come in a variety of geometries, but the so-called 'trouser-tear' test, shown schematically in Figure 2.33, has several advantages when compared to other test specimen configurations.

The primary advantage is the simple analysis of the specimen that leads to a healing efficiency calculated by simply dividing the average tear force of the

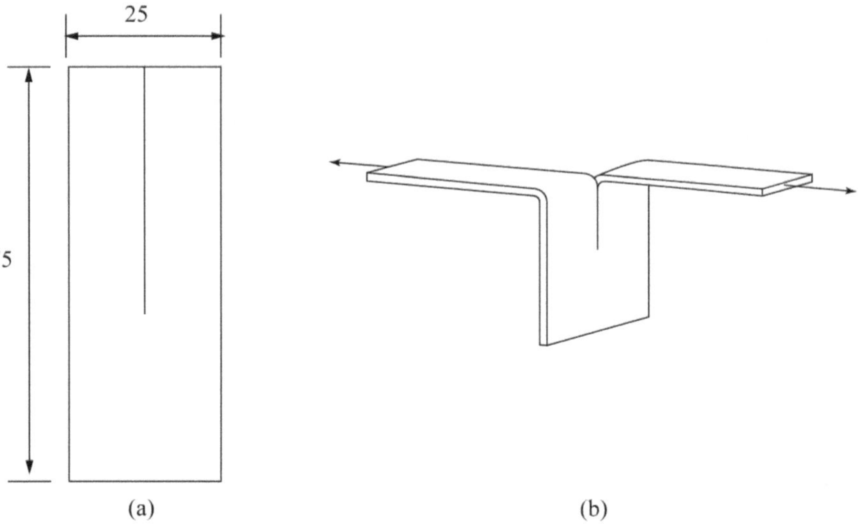

Figure 2.33 Tear specimen geometry (a) and schematic of a loaded tear specimen during testing (b).

initial test by the average tear force of the healed test. Based on this tear test, the elastomer developed by Keller, White, and Sottos demonstrated high healing efficiencies and was capable of fully recovering the virgin tear strength under certain circumstances.[47] The full recovery of tear strength was surprising, and unexpected. These high levels of healing were attributed to local areas of strong adhesion between the healing chemistry-derived polymer and the surrounding matrix. As the tear approached a region with these properties, the tear would deviate into surrounding virgin material and proceed along a new path (see Figure 2.34(b)). Furthermore, the addition of microcapsules improved tear properties for the virgin material, see Figure 2.34(a).

2.4.4 Restoration of Interfacial Strength

In addition to the bulk material healing described above, micro-debond specimens were utilized by Blaiszik and co-workers to investigate the healing of interfacial bonding in a model composite system.[92] The micro debond specimen is a modified fiber-pull out test that attempts to measure the interfacial shear strength (IFSS) between an individual fiber/matrix system.[93–95] A sample is manufactured by placing a small drop of epoxy on or other matrix material on a single fiber, which is cured between two microscope slides. The fiber is then pulled out of the matrix drop. In the self-healing studies, the DCPD/Grubbs system[2] was deposited on the fiber/matrix interface as shown in Figure 2.35.

An alternative approach to the investigation of interfacial healing was presented by Sandana, Itaya, and Shindo.[96,97] Interfacial healing specimens

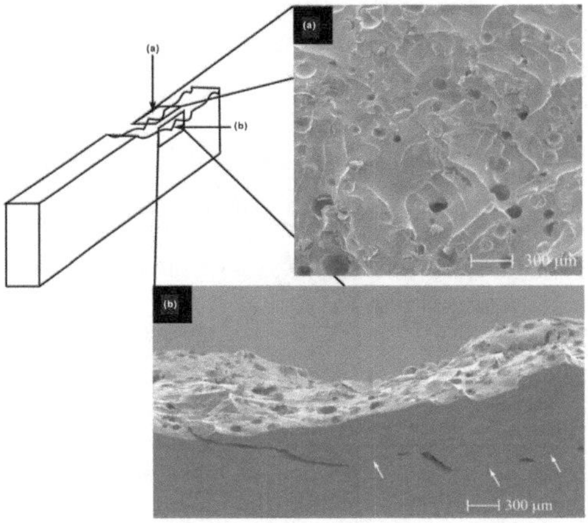

Figure 2.34 SEM images of a tear surface (a) and a deviated tear (b) for a self-healing PDMS elastomer. Reproduced from ref. 47 with permission from John Wiley & Sons, Copyright 2007 Wiley-VCH Verlag GmbH & Co. KGaA, Weinheim.

were created by coating fibers with a slurry of DCPD microcapsules and Grubbs catalyst in an epoxy resin that was starved of curing agent. The resulting fibers were then embedded within a tensile bar and loaded until failure. The separated halves were put back into contact and allowed to heal, and then loaded again until failure (Figure 2.36). Healing efficiencies of approximately 20% were achieved using this indirect test method.

2.5 Applications of Microcapsule Healable Materials in Real World Applications

It is clear from the examples selected in this chapter that research into healable polymers containing microcapsules has progressed significantly since the seminal publication by White and Sottos in 2001.[2] In order to impact on society, however, these materials have to demonstrate the possibility of solving real world problems. Although healable materials containing microcapsules have yet to achieve commercialization, laboratory scale prototyping has begun to point the way towards markets that would benefit from their unique properties, as shown in the select examples set out in the following sections below.

2.5.1 Healable Polymeric Materials Containing Microcapsules for Anti-corrosion Coatings

Self-healing barriers have a long history in the paint and coatings industry, but self-healing in this context is not strictly the same property as described in this chapter. Most 'self-healing' coatings take advantage of low T_g flow and creep to slowly level a scratch or other physical defect in the coating.

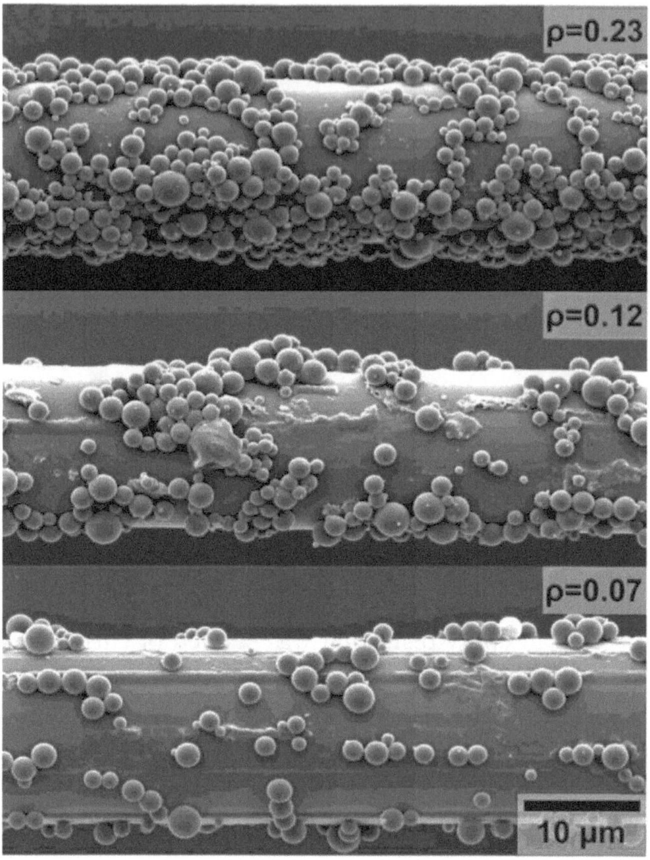

Figure 2.35 Nanocapsules adhered to the surface of a glass fiber. Reproduced from ref. 92 with permission from John Wiley & Sons, Copyright 2010 Wiley-VCH Verlag GmbH & Co. KGaA, Weinheim.

Using the definition adopted in this chapter, a self-healing coating must possess a repair functionality that is triggered by mechanical damage, just as in the other cases described above. The recovery of barrier properties is frequently assessed in a qualitative manner, such as the recovery of corrosion resistance or in a 'go/no-go' recovery metric as in the case of pressure containing structures, such as fuel tanks. Qualitative testing of such systems can be performed using linear voltammetry on coatings that are non-conductive.

2.5.1.1 Corrosion Barrier/Paint Healing

The first demonstration of autonomic recovery of barrier properties utilized a siloxane-based healing chemistry and investigated the self-repair of a coating on a steel substrate.[46] As shown in Figure 2.37(a), a self-healing

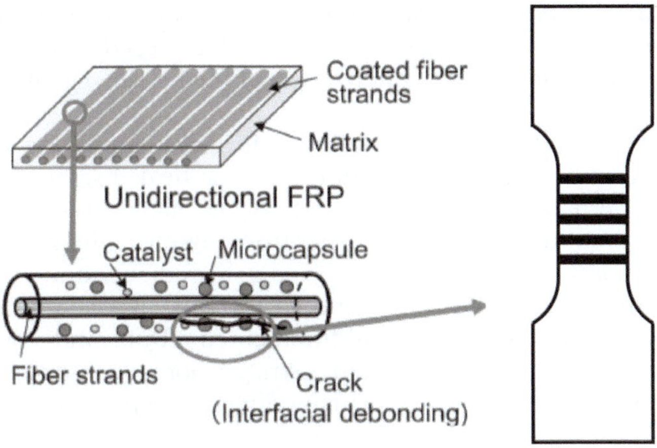

Figure 2.36 Tensile-based interfacial healing system. Adapted from ref. 97 with permission from Sage, Copyright 2006.

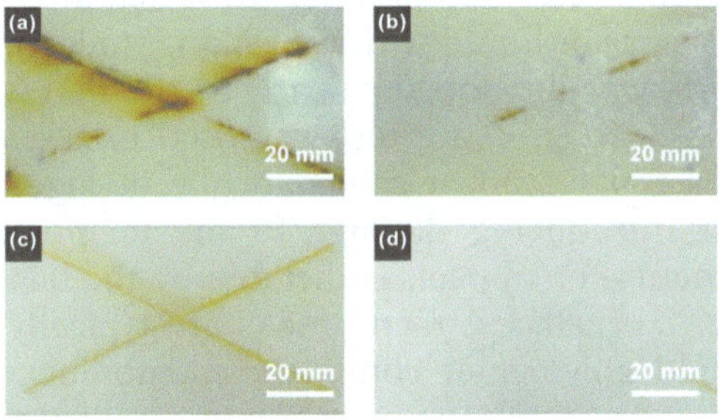

Figure 2.37 Recovery of corrosion resistance properties of a self-healing coating. Part (a) and (c) represent control specimens with no self-healing functionality. Images (b) and (c) are self-healing coating materials. The top images are a custom-made coating and the bottom images are commercial marine paints modified to become self-healing. Reproduced from ref. 46 with permission from John Wiley & Sons, Copyright 2009 Wiley-VCH Verlag GmbH & Co. KGaA, Weinheim.

coating that has been damaged using a scratch can recover complete barrier properties. The samples in Figure 2.37 were scribed using a commercial coating scriber, prior to resting for a 24 hour long healing period and were then immersed in salt water. The recovery of barrier properties was found to be complete when the self-healing material was added to a commercial marine paint (Figure 2.37(d)). The incomplete recovery in a laboratory-made coating, Figure 2.37(b), is primarily the result of the poor scratch resistance of the base material – compare Figure 2.37(a) to (c).

Since this first demonstration of film repair, there have been several additional reports described in the literature. Huang and co-workers used a microencapsulated diisocyanate embedded in an epoxy film to repair scratch damage.[98] Since this system was based on isocyanate chemistry, this exploited one-component healing chemistries in that it used ambient moisture to initiate and complete the polymerization of the healing agent released by the crack.

In the only demonstration of healing using electrospinning as the microencapsulation procedure, Park and Braun also healed scratch damage in a coating.[23] The polymer was a UV-cured polyurethane acrylate, which was required to prevent the PVP shell polymer from dissolving in the matrix. PVP is soluble in most coating polymers such as polyurethanes and epoxides but the rapid curing of the UV coating limited the time available for shell dissolution. Healing chemistry was addition-cured PDMS and complete recovery of barrier properties was demonstrated both qualitatively through visual analysis and quantitatively by using linear sweep voltammetry.

2.5.2 Healable Polymeric Materials Containing Microcapsules: Pressure Barrier Healing

There are a multitude of everyday products in which a polymeric barrier is required to enclose a gas or liquid without leaking. Indeed, the concept of a self-healing fuel tank was developed as early as the first world war. In two such reports, a one-sided pressure test was used to assess the healing of damage. The test setup is shown schematically in Figure 2.38. A test is performed by first introducing damage to the membrane. The type of damage, either puncture or impact, is dictated by the base material of the specimen. Flexible, fabric-based membranes are typically punctured by a needle, whereas rigid polymer or composite membranes are impacted or indented. After being damaged, the membranes are allowed to heal and are then placed in the pressure cell. Pressure is applied to one side of the membrane and the leakage pressure is measured on the opposite side. Healing efficiency is calculated either as a 'leak/no leak' statistic based on a large number of samples or, potentially, as a reduction in gas flow rate as a result of the repair.

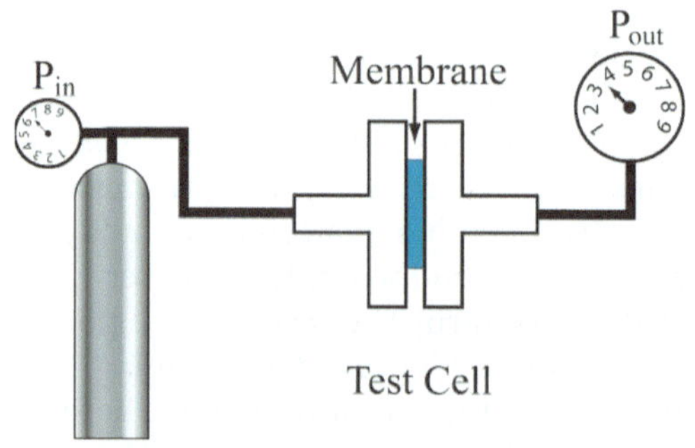

Figure 2.38 Schematic of the membrane/pressure support test cell.

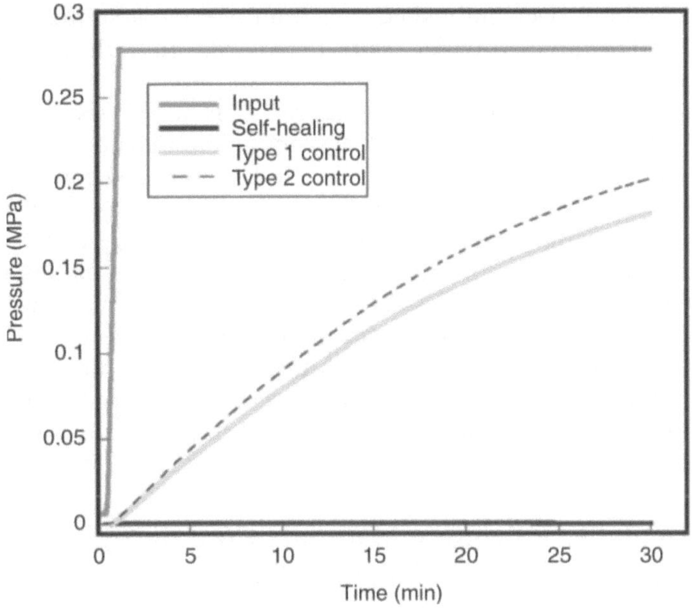

Figure 2.39 Pressure *vs.* time traces for two non-healing controls (Type 1 and Type 2) and a fully self-healing material compared to the input pressure trace. Reproduced from ref. 100 with permission from Sage Publications, Copyright 2010.

In the first report of an autonomic self-healing membrane, a flexible membrane based on siloxane chemistry was fabricated and demonstrated. This material consisted of a PDMS-based layer laminated between two high-performance barrier layers. Damage was introduced using a needle puncture to simulate pinhole damage, as this system was intended for use in an inflatable space habitat.[98] Repair of this type of damage is of particular interest for the feasibility of composite-based fuel tanks. Moll and co-workers studied the recovery of pressure-holding capacity for a plain weave E-glass/epoxy composite. Square sample coupons were damaged with a sharp indenter, allowed a room temperature healing period, and then tested in a pressure cell similar to the one shown in Figure 2.38.[99,100] Results from this testing are shown in Figure 2.39, which show that the self-healing material was able to completely restore the pressure carrying capacity for the material when compared to the non-healing control specimens.

2.5.3 Restoration of Conductivity

The restoration or repair of conductivity is a relatively new area for self-healing materials, however, such material capabilities could find application across a broad range of electronic devices to fortuitously extend their life-spans and also reduce waste. Reports of the autonomic restoration of conductivity have been published. For example one system used a microencapsulated charge transfer salt TTF–TCNQ (tetrathiafulvalene–tetracyanoquinodimethane) in a two-capsule formulation that formed the conductive salt when each part was mixed.[101] Healing was tested by fabricating a circuit with a pre-made flaw using sputter-coated gold on a glass slide. Capsules were crushed on the slide and the released contents formed a conductive path from one side to the other. The resistance of the healed circuit was measured and found to slowly decrease the resistance of the circuit as time progressed. An alternative approach was also developed by the same group using capsules filled with a solvent, hexyl acetate. For this self-healing system, conductive silver ink was used to create a circuit. A cut in a thin polymer coating over the conductive traces simultaneously broke the conductive path and released the solvent healing agent. The released solvent dissolved the polymer carrier in the silver paint, causing the paint to flow, and slowly restored conductivity over a week to the initial level.[102]

2.5.4 Healable Polymeric Materials Containing Microcapsules for Dental Fillings

It is clear from the examples in this chapter that fabrication of microcapsules containing polymeric materials is significantly more complex than producing typical bulk polymeric materials. This has cost implications for products making use of this technology and makes it likely that initial areas for commercialization will be for low volume, high value applications. As such, the market for a new generation of dental fillings has become an active target for the commercialization of this class of healing material.[103]

Crosslinked polymeric composite resins have become the standard dental filling over the past 20 years. However, as a consequence of the regular applied force to the filling whilst eating, and the large variations in temperature that can be experienced in the mouth (from ice cold drinks to hot coffee), the resins are susceptible to microcracking and ultimately failure, making them an ideal target for this new technology. However, achieving this goal is complicated by the specific needs of the material to: (i) be easily applied to the tooth cavity; (ii) exhibit the appropriate curing time; (iii) have the necessary structural longevity. In addition, these design criteria must be achieved whilst using only demonstrably biocompatible materials.

In 2016, Wu *et al.* produced UF microcapsules containing the monomer triethylene glycol dimethacrylate (TEGDMA) and *N*,*N*-dihydroxyethyl-*p*-toluidine (DHEPT) which is a biocompatible accelerator for the free radical polymerization of TEGDMA (Figure 2.40).[104] The components of the healing solution were encapsulated in microcapsules using the standard oil-in water emulsion technique. Once the microcapsules were isolated they were incorporated into typical dental filling formulations (glycidyl dimethacrylate (GDM) and TEGDMA, Figure 2.39) which contained a photo initiator and benzoyl peroxide (BPO) (0.5 wt%) to act as the self-healing initiator. This

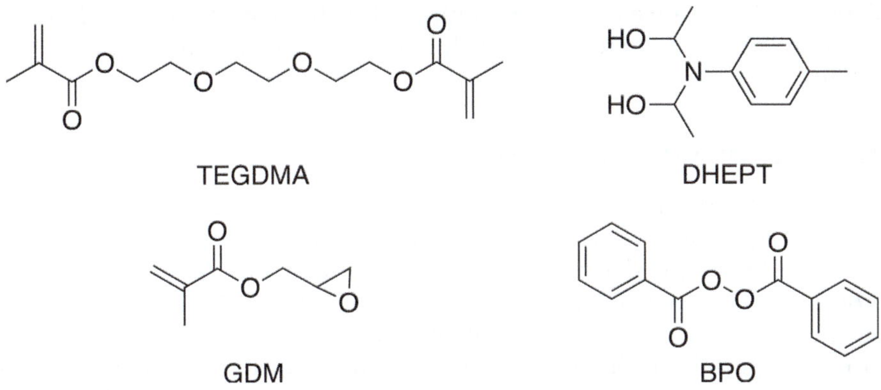

Figure 2.40 Structures of TEGDMA, DHEPT, GDM and BPO.

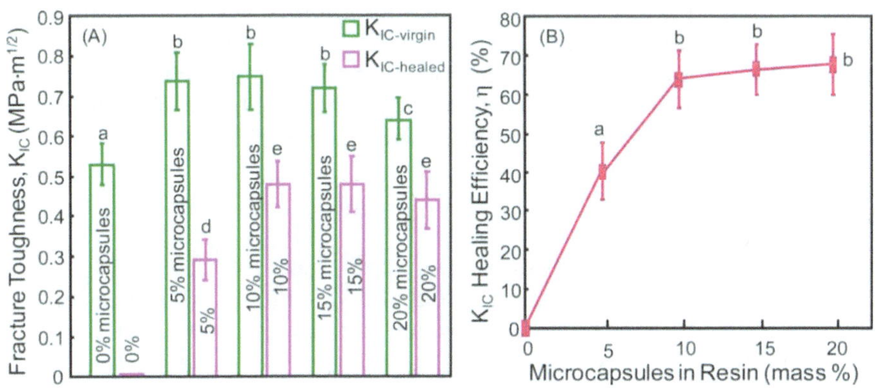

Figure 2.41 (A) Virgin (green) and healed (pink) fracture toughness of sample dental fillings containing between 0 and 20% microcapsules. (B) Healing efficiency for sample dental fillings containing between 0 and 20% microcapsules. Reproduced from ref. 104 with permission from Elsevier, Copyright 2015.

suspension was irradiated to produce the solid microcapsule containing sample fillings for testing.

Sample dental fillings containing between 0 to 20 wt% microcapsules were produced and subject to a fracture-heal-fracture cycle whilst measuring the physical properties of the materials during each fracture. The polymer lacking microcapsules (0 wt%) had a fracture toughness of approximately 53 MPa m$^{1/2}$ and showed no healing ability under the testing conditions. However, samples containing between 5 and 20% microcapsules were found to exhibit a high fracture toughness (up to 75 MPa m$^{1/2}$) compared to the standard sample Figure 2.41(A) and healing efficiencies of approximately 65% for samples containing 10 to 20% microcapsules Figure 2.41(B). These data suggest that fillings containing the microcapsule technology could offer significantly enhanced lifetimes compared to the current standard, which would result in a dramatic reduction in cost and inconvenience to the patient.

2.6 Summary and Outlook

There has been rapid progress in the development of microcapsule-based self-healing materials in the 20 years since the first literature reports. Available healing chemistries have expanded and the microencapsulation procedures have become more varied and sophisticated. The general outlook for this class of materials is extremely promising in a wide variety of applications, especially in the structural composites industry. Commercialization efforts are underway and it is anticipated that a consumer or industrial-based product will be introduced in the near future, most likely a coating or paint. Microcapsule-based healing is continuing to develop with papers being published regularly on new healing chemistries and formulations. One of the most critical areas of research that is left is the incorporation of a function that would provide autonomic confirmation of healing. Ideally, any self-healing material would also come with the capability of informing an inspector that healing has been successfully accomplished.

References

1. R. P. Wool and K. M. O'Conner, *J. Appl. Phys.*, 1981, **52**, 5953.
2. S. R. White, N. R. Sottos, P. H. Geubelle, J. S. Moore, M. R. Kessler, S. R. Sriram, E. N. Brown and S. Viswanathan, *Nature*, 2001, **409**, 794.
3. S. Freitas, H. P. Merkle and B. Gander, *J. Controlled Release*, 2005, **102**, 313.
4. C. Thies, *Drugs Pharm. Sci.*, 1996, **73**, 1.
5. E. Brown, M. Kessler, N. Sottos and S. White, *J. Microencapsulation*, 2003, **20**, 719.
6. X. Liu, J. K. Le and M. R. Kessler, *Macromol. Res.*, 2011, **19**, 1056.
7. B. Blaiszik, N. Sottos and S. White, *Compos. Sci. Technol.*, 2008, **68**, 978.
8. D. S. Xiao, Y. C. Yuan, M. Z. Rong and M. Q. Zhang, *Polymer*, 2009, **50**, 560.
9. T. Nesterova, K. Dam-Johansen and S. Kiil, *Prog. Org. Coat.*, 2011, **70**, 345.
10. A. C. Jackson, J. A. Bartelt, K. Marczewski, N. R. Sottos and P. V. Braun, *Macromol. Rapid Commun.*, 2011, **32**, 82.
11. S. Cho, H. Andersson, S. White, N. Sottos and P. Braun, *Adv. Mater.*, 2006, **18**, 997.
12. M. Jacquemond, N. Jeckelmann, L. Ouali and O. P. Haefliger, *J. Appl. Polym. Sci.*, 2009, **114**, 3074.
13. D. A. McIlroy, B. J. Blaiszik, M. M. Caruso, S. R. White, J. S. Moore and N. R. Sottos, *Macromolecules*, 2010, **43**, 1855.
14. J. Yang, M. W. Keller, J. S. Moore, S. R. White and N. R. Sottos, *Macromolecules*, 2008, **41**, 9650.
15. M. Huang and J. Yang, *J. Mater. Chem.*, 2011, **21**, 11123.

16. M. M. Caruso, B. J. Blaiszik, H. Jin, S. R. Schelkopf, D. S. Stradley, N. R. Sottos, S. R. White and J. S. Moore, *ACS Appl. Mater. Interfaces*, 2011, **2**, 1195.
17. S. Kang, M. Baginska, S. R. White and N. R. Sottos, *ACS Appl. Mater. Interfaces*, 2015, **7**, 10952.
18. H. Yi, Y. Yang, X. Gu, J. Huang and C. Wang, *J. Mater. Chem. A*, 2015 **3**, 13749.
19. C. Dry, *Compos. Struct.*, 1996, **35**, 263.
20. J. W. C. Pang and I. P. Bond, *Compos. Sci. Technol.*, 2005, **65**, 1791.
21. S. D. Mookhoek, H. R. Fischer and S. van der Zwaag, *Composites, Part A*, 2012, **43**, 2176.
22. M. M. Caruso, D. A. Delafuente, V. Ho, N. R. Sottos, J. S. Moore and S. R. White, *Macromolecules*, 2007, **40**, 8830.
23. J.-H. Park and P. V. Braun, *Adv. Mater.*, 2010, **22**, 496.
24. S. Sinha-Ray, D. D. Pelot, Z. P. Zhou, A. Rahman, X. F. Wu and A. L. Yarin, *J. Mater. Chem.*, 2012, **22**, 9138.
25. J. J. Vericella, *et al.*, *Nat. Commun.*, 2015, **6**, 1.
26. A. S. Utada, E. Lorenceau, D. R. Link, P. D. Kaplan, H. A. Stone and D. A. Weitz, *Science*, 2005, **308**, 537–541.
27. Y. Geng, S. Ling, J. Huang and J. Xu, *Small*, 2020, **16**, 1906357.
28. P. Zhu, P. Zhu, L. Wang and L. Wang, *Lab Chip*, 2017, **17**, 34.
29. E. Dressaire and A. Sauret, *Soft Matter*, 2016, **13**, 37.
30. D. Serp, E. Cantana, C. Heinzen, U. V. Stockar and I. W. Marison, *Biotechnol. Bioeng.*, 2000, **70**, 41.
31. M. Fujiwara, K. Shiokawa, Y. Tanaka and Y. Nakahara, *Chem. Mater.*, 2004, **16**, 5420.
32. R. Atkin, P. Davies, J. Hardy and B. Vincent, *Macromolecules*, 2004, **21**, 7979.
33. K. Bean, C. F. Black, N. Govan, P. Reynolds and M. R. Sambrook, *J. Colloid Interface Sci.*, 2012, **366**, 16.
34. B. Blaiszik, M. Caruso, D. Mcilroy, J. S. Moore, S. White and N. Sottos, *Polymer*, 2009, **50**, 990.
35. X. Liu, J. K. Lee, S. H. Yoon and M. R. Kessler, *J. Appl. Polym. Sci.*, 2006, **101**, 1266.
36. L. M. Meng, Y. C. Yuan, M. Z. Rong and M. Q. Zhang, *J. Mater. Chem.*, 2010, **20**, 6030.
37. M. W. Keller and N. R. Sottos, *Exp. Mech.*, 2006, **46**, 725.
38. J. D. Rule and J. S. Moore, *Macromolecules*, 2002, **35**, 7878.
39. M. R. Kessler and S. R. White, *J. Polym. Sci., Part A: Polym. Chem.*, 2002, **40**, 2373.
40. X. Liu, X. Sheng, J. K. Lee, M. R. Kessler and J. S. Kim, *Compos. Sci. Technol.*, 2009, **69**, 2102.
41. G. O. Wilson, M. M. Caruso, N. T. Reimer, S. R. White, N. R. Sottos and J. S. Moore, *Chem. Mater.*, 2008, **20**, 3288.
42. A. S. Jones, J. D. Rule, J. S. Moore, S. R. White and N. R. Sottos, *Chem. Mater.*, 2006, **18**, 1312.
43. T. C. Mauldin and M. R. Kessler, *J. Mater. Chem.*, 2010, **20**, 4198.

44. J. Rule, E. Brown, N. Sottos, S. White and J. Moore, *Adv. Mater.*, 2005, **17**, 205.
45. J. M. Kamphaus, J. D. Rule, J. S. Moore, N. R. Sottos and S. R. White, *J. R .Soc., Interface*, 2008, **5**, 95.
46. S. H. Cho, S. R. White and P. V. Braun, *Adv. Mater.*, 2009, **27**, 645.
47. M. W. Keller, S. R. White and N. R. Sottos, *Adv. Funct. Mater.*, 2007, **17**, 2399.
48. M. M. Caruso, B. J. Blaiszik, S. R. White, N. R. Sottos and J. S. Moore, *Adv. Funct. Mater.*, 2008, **18**, 1898.
49. G. O. Wilson, J. W. Henderson, M. M. Caruso, B. J. Blaiszik, P. J. Mcintire, N. R. Sottos, S. R. White and J. S. Moore, *J. Polym. Sci., Part A: Polym. Chem.*, 2010, **48**, 2698.
50. H. P. Wang, Y. C. Yuan, M. Z. Rong and M. Q. Zhang, *Macromolecules*, 2010, **43**, 595.
51. L. Yao, Y. C. Yuan, M. Z. Rong and M. Q. Zhang, *Polymer*, 2011, **52**, 3137.
52. M. Zako and N. Takano, *J. Intell. Mater. Syst. Struct.*, 1999, **10**, 836.
53. I. P. Bond, R. S. Trask and H. R. Williams, *MRS Bull.*, 2008, **33**, 770.
54. J. Pang and I. Bond, *Composites, Part A*, 2005, **36**, 183.
55. R. S. Trask and I. P. Bond, *Smart Mater. Struct.*, 2006, **15**, 704.
56. R. S. Trask, I. P. Bond and C. O. A. Semprimoschnig, European Space Agency, Noordwijk, 2200 AG, Netherlands, Collioure, France, 2006, 8.
57. G. Williams, R. Trask and I. Bond, *Composites, Part A*, 2007, **38**, 1525.
58. H. R. Williams, R. S. Trask and I. P. Bond, *Smart Mater. Struct.*, 2007 **16**, 1198.
59. T. Yin, M. Rong, M. Zhang and G. Yang, *Compos. Sci. Technol.*, 2007, **67**, 201.
60. T. S. Coope, U. F. J. Mayer, D. F. Wass, R. S. Trask and I. P. Bond, *Adv. Funct. Mater.*, 2011, **21**, 4624.
61. C. Y. Yan, Z. R. Min, Q. Z. Ming, J. Chen, C. Y. Gui and M. L. Xue, *Macromolecules*, 2008, **41**, 5197.
62. H. Jin, G. M. Miller, N. R. Sottos and S. R. White, *Polymer*, 2011, **52**, 1628.
63. H. C. Kolb, M. G. Finn and K. B. Sharpless, *Angew. Chem., Int. Ed.*, 2001, **40**, 2004.
64. M. Gragert, M. Schunack and W. H. Binder, *Macromol. Rapid Commun.*, 2011, **32**, 419.
65. M. M. Caruso, D. A. Delafuante, V. Ho, N. R. Sottos, J. S. Moore and S. R. White, *Macromolecules*, 2007, **40**, 8830.
66. A. C. Jackson, J. A. Bartelt and P. V. Braun, *Adv. Funct. Mater.*, 2011, **21**, 4705.
67. S. D. Mookhoek, S. C. Mayo, A. E. Hughes, S. A. Furman, H. R. Fischer and D. Z. S. van, *Adv. Eng. Mater.*, 2010, **12**, 228.
68. G. V. Kolmakov, R. Revanur, R. Tangirala, T. Emrick, T. P. Russell, A. J. Crosby and A. C. Balazs, *ACS Nano*, 2010, **4**, 1115.

69. M. R. Kessler, N. R. Sottos and S. R. White, *Composites, Part A*, 2003, **34**, 743.
70. M. R. Kessler and S. R. White, *Composites, Part A*, 2001, **32**, 683.
71. E. N. Brown, N. R. Sottos and S. R. White, *Exp. Mech.*, 2002, **42**, 372.
72. Y. C. Yuan, Y. Ye, M. Z. Rong, H. Chen, J. Wu, M. Q. Zhang, S. X. Qin and G. C. Yang, *Smart Mater. Struct.*, 2010, **20**, 015024.
73. S. M. Bleay, C. B. Loader, V. J. Hawyes, L. Humberstone and P. T. Curtis, *Composites, Part A*, 2001, **32**, 1767.
74. S. Mostovoy, P. B. Crosley and E. J. Ripling, *J. Mater.*, 1967, **2**, 661.
75. D. Broek, *Elementary Engineering Fracture Mechanics*, Martinus Nijhoff Publishers, Boston, 1986.
76. Y. C. Yuan, M. Z. Rong, M. Q. Zhang and G. C. Yang, *Polymer*, 2009, **50**, 5771.
77. E. N. Brown, S. R. White and N. R. Sottos, *J. Mater. Sci.*, 2004, **39**, 1703.
78. G. Li, G. Ji and O. Zhenyu, *Int. J. Adhes. Adhes.*, 2012, **35**, 59.
79. J. D. Rule, N. R. Sottos and S. R. White, *Polymer*, 2007, **48**, 3520.
80. E. L. Kirkby, J. D. Rule, V. J. Michaud, N. R. Sottos, S. R. White and J.-A. E. Manson, *Adv. Funct. Mater.*, 2008, **18**, 2253.
81. C. L. Mangun, A. C. Mader, N. R. Sottos and S. R. White, *Polymer*, 2010, **51**, 4063.
82. G. O. Wilson, J. S. Moore, S. R. White, N. R. Sottos and H. M. Andersson, *Adv. Funct. Mater.*, 2008, **18**, 44.
83. E. Brown, S. White and N. Sottos, *Compos. Sci. Technol.*, 2005, **65**, 2466.
84. J. L. Tzou, C. H. Hsueh, A. G. Evans and R. O. Ritchie, *Acta Metall. Mater.*, 1985, **33**, 117.
85. J. L. Tzou, S. Suresh and R. O. Ritchie, *Acta Metall. Mater.*, 1985, **33**, 105.
86. E. Brown, S. White and N. Sottos, *Compos. Sci. Technol.*, 2005, **65**, 2474.
87. A. S. Jones, J. D. Rule, J. S. Moore, N. R. Sottos and S. R. White, *J. R .Soc., Interface*, 2007, **4**, 395.
88. M. W. Keller, S. R. White and N. R. Sottos, *Polymer*, 2008, **49**, 3136.
89. A. J. Patel, N. R. Sottos, E. D. Wetzel and S. R. White, *Composites, Part A*, 2010, **41**, 360.
90. L. Yao, M. Z. Rong, M. Q. Zhang and Y. C. Yuan, *J. Mater. Chem.*, 2011, **21**, 9060.
91. S. A. Hayes, F. R. Jones, K. Marshiya and W. Zhang, *Composites, Part A*, 2007, **38**, 1116.
92. B. J. Blaiszik, B. A. Baginska, S. R. White and N. R. Sottos, *Adv. Funct. Mater.*, 2010, **20**, 3547.
93. S. Zhandarov, Y. Gorbatkina and E. Mader, *Compos. Sci. Technol.*, 2006, **66**, 2610.
94. S. Zhandarov and E. Mader, *J. Adhes. Sci. Technol.*, 2005, **19**, 817.
95. S. Zhandarov and E. Mader, *Compos. Sci. Technol.*, 2005, **65**, 149.
96. K. Sanada, N. Itaya and Y. Shindo, *Open Mech. Eng. J.*, 2008, **2**, 97.
97. K. Sanada, I. Yasuda and Y. Shindo, *Plast., Rubber Compos.*, 2007, **35**, 67.
98. G. C. Huang, J. K. Lee and M. R. Kessler, *Macromol. Mater. Eng.*, 2011, **296**, 965.

99. B. A. Beiermann, M. W. Keller and N. R. Sottos, *Smart Mater. Struct.*, 2009, **18**, 085001.
100. J. L. Moll, S. R. White and N. R. Sottos, *J. Compos. Mater.*, 2010, **44**, 2573.
101. S. A. Odom, M. M. Caruso, A. D. Finke, A. M. Prokup, J. A. Ritchey, J. H. Leonard, S. R. White, N. R. Sottos and J. S. Moore, *Adv. Funct. Mater.*, 2010, **20**, 1721.
102. S. A. Odom, S. Chayanupatkul, B. J. Blaiszik, O. Zhao, A. C. Jackson, P. V. Braun, N. R. Sottos, S. R. White and J. S. Moore, *Adv. Mater.*, 2012, **24**, 2578.
103. For a review of this area see: X. Wang and T. Ding, *J. Funct. Biomater.*, 2024, **15**, 165.
104. J. Wu, M. D. Weir, Q. Zhang, C. Zhou, M. A. Melo and H. H. Xu, *Dent. Mater.*, 2016, **32**, 294.

CHAPTER 3

Reversible Covalent Bond Formation as a Strategy for Healable Polymer Networks

CHRISTOPHER J. KLOXIN

Department of Materials Science and Engineering & Department of Chemical and Biomolecular Engineering, University of Delaware, 150 Academy Street, Newark, DE 19716 USA
Email: cjk@udel.edu

3.1 Introduction

The most sustainable material is one that does not need to be replaced. While recycling is an essential aspect of sustainability, especially in managing single-use polymeric materials, it poses significant infrastructural and processing challenges. These processes consume substantial energy, further impacting the environment. Alternatively, mending or healing polymeric materials offers a practical solution to enhance their sustainability.

Polymeric materials are typically categorized as either thermoplastics or thermosets, depending on whether the material is covalently crosslinked. Thermoplastics can be remolded, mended, or recycled, whereas thermosets have permanent crosslinks and are essentially 'set' once fabricated. While thermosets are often chemically and structurally robust, these same attributes are incompatible with most reprocessing operations. Once the underlying polymer structure degrades due to fatigue or environmental stress, few mechanisms exist to prevent material failure. Consequently, the sustainability challenge in thermosets has driven innovations in their chemical designs.

Polymer Chemistry Series No. 41
Healable Polymer Systems: Fundamentals, Synthesis and Applications, 2nd Edition
Edited by Barnaby W. Greenland and Wayne Hayes

Published by the Royal Society of Chemistry, www.rsc.org

Since the previous edition of this chapter ten years ago, polymer networks with reversible covalent crosslinks have emerged as a distinct class of material. The terms 'covalent adaptable networks' (CANs) and 'vitrimers', were first used in 2010[1] and 2012[2] respectively, and are now commonly used in the literature. The incorporation of reversible covalent linkages enables the network strands to undergo bond shuffling and rearrangement (see Figure 3.1). These linkages impart unique properties that blend characteristics of both thermosets and thermoplastics. Among these properties is the ability of repair at a molecular level, designated here as 'healing'. While there are numerous potential applications for CANs, perhaps the most compelling is their potential to create innovative sustainable materials to address the ever-growing problem of polymer waste production.

Incorporating reversible covalent crosslinks into a polymer network offers a multifaceted approach to extending the service lifetime of a material. This

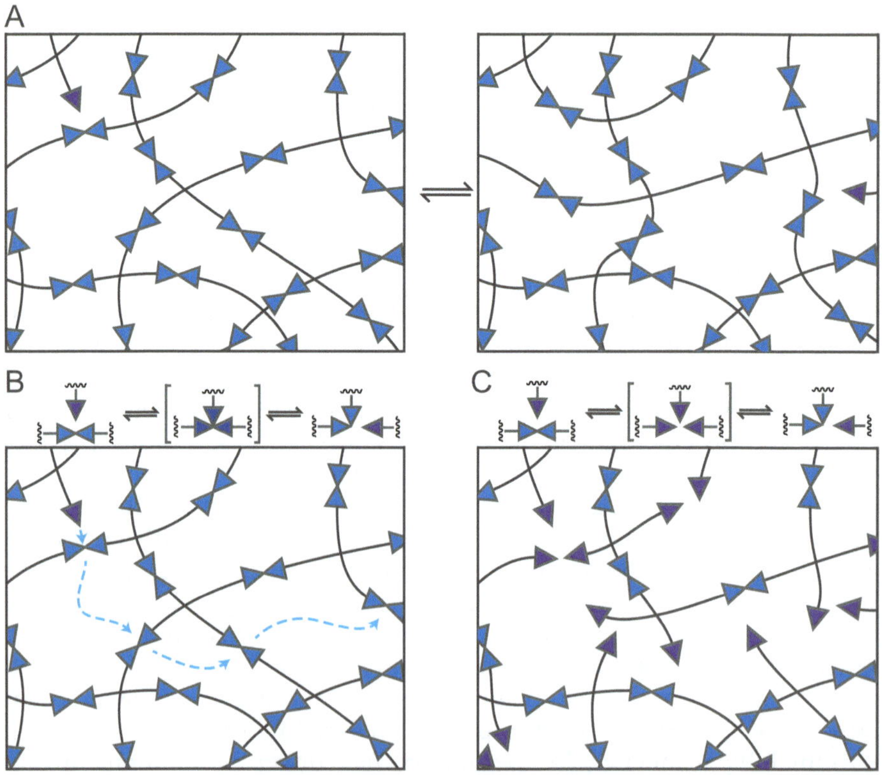

Figure 3.1 Illustration of network strand rearrangement within covalent adaptable networks (CANs), where bowties represent reversible covalent crosslinks and triangles represent reactive intermediates. Two potential topological states of the CAN where the number of reversible crosslinks is the same (A). The transition between the two states can follow either an associative (B) or a dissociative (C) mechanism, depending on the type of reversible covalent crosslink.

can range from subtle microscopic bond rearrangement that reduces defects to macroscopic flow that repairs larger-scale crazes and cracks. However, not all implementations of reversible covalent chemistry result in the same healing outcome. The underlying mechanism of the reversible covalent chemistry (*e.g.*, see Figure 3.1(B) and (C)) significantly impacts the macroscopic properties of the network. When designing a CAN for specific applications, it is crucial to consider the network topology, the placement of reversible covalent crosslinks within the network, and the type of reversible covalent chemistry used.

This chapter will highlight the key design features of CANs and explain the underlying mechanisms that lead to various healing phenomena. Several recent (and not so recent) reversible crosslinking chemistries will be discussed. This is not intended as a comprehensive review, partly because several excellent reviews already exist,[3–12] but mainly because many of the reversible crosslinks are deeply rooted in a wide range of chemistries dating back several decades. Many of the recent reports on CANs are a reapplication of old materials and reactions. This is not to imply these materials are trivial but rather that there is a renewed appreciation for their complex nature, making them well-suited for many modern and specialized applications. This chapter will place recent advances in CANs within the broader context of reversible covalent chemistries, outlining the key CAN design principles that will enable their widespread implementation in healing applications and provide a toolbox for developing new CANs.

3.2 A Brief Historical Account

The concept that covalently crosslinked networks are capable of non-destructive creep has been well-known and extensively studied since the 1930s and even earlier, particularly in vulcanized rubber.[13] In these materials, polymer chains are crosslinked by sulfur bridges, which can undergo scission and further crosslinking. A phenomenological description is used where an elastomeric material is elongated to a prescribed strain, and the material's equilibrium length is measured over time. The change in material dimensions is attributed to creep, facilitated by the breaking and reforming of bonds.

Originally, the stress relaxation and creep observed in vulcanized rubbers were attributed to strain-induced polymer degradation; however, Green and Tobolsky[14] reasoned that the bonds must be continually breaking and reforming, based on the observation that relaxation rates were independent of strain (for small extensions). Moreover, the bond-breaking and reforming processes occur at equal rates and do not contribute to the stress for small deformations, suggesting these crosslinking reactions are at equilibrium. It is important to note that creep in crosslinked polymers was considered a defect at the time, and most efforts focused on reducing this phenomenon or developing new materials that lacked this quality.

In the following decades, there was significant interest in covalent crosslinked materials with thermo-reversible properties. It was already well-known that many polymerizations had a ceiling temperature, which, when reached, would reverse the material to a lower molecular weight (and even to its monomer constituents). Unfortunately, these high temperatures are often accompanied by irreversible side reactions or decomposition, limiting their use as a means of reprocessing the polymer material. Indeed, the reduction of side reactions remains a major challenge in the development of thermo-reversible covalent adaptable networks today.

In 1966, James Craven of the DuPont Company introduced a thermo-reversible crosslinking material utilizing the reversibility of the [4 + 2] cycloaddition reaction, also known as the Diels–Alder reaction. In his 1969 patent, he wrote, "Polymer products should, for most uses, be insoluble in common solvents. For many purposes, such products should also have the ability to [be] post-formed... Heretofore, these properties were considered to be mutually exclusive".[15] This sentiment essentially captures the motivation for developing processable covalent networks at the time, as well the current development of new covalent adaptable networks.

The development of reversible or dynamic covalent chemistry (DCC)[4–6] in several small molecule (*e.g.* DCL[4]) and supramolecular chemistry[6,16] applications has ushered in the era of covalent adaptable networks. Incorporating these chemistries into covalently crosslinked networks imparts the reversible character that defines covalent adaptable networks. The resulting properties of these networks are not only dependent on the kinetic and thermodynamic properties of the reversible covalent linkages but also on the network design and connectivity. The subsequent section explores the various design elements that must be considered to create novel covalent adaptable networks for healing applications.

3.3 Covalent Adaptable Network (CAN) Design

The design of a covalent adaptable network (CAN) must take into account two main considerations: (1) network topology and (2) the type of reversible chemistry. The constituent monomers and the network formation mechanism determine the location and distribution of reversible crosslinks. The type of reversible crosslinks determines how network strands rearrange on the microscopic scale. The network topology, combined with the type of reversible crosslinks, ultimately dictates how efficiently the material will heal or relax on a macroscopic scale.

3.3.1 Network Topology

Covalently crosslinked networks are phenomenologically characterized by the presence of a zero-frequency (long-time) modulus, which represents the network's capacity to indefinitely store energy in the form of elastic deformation. The macroscopic stress is transmitted through network backbone strands, which are deformed into higher-entropy conformational states. In CANs, the ultimate dissipation of this stored energy occurs through the breaking and reforming of elastically active strands within the network. Therefore, the placement of dynamic covalent crosslinks within the network significantly affects stress relaxation. Specifically, CANs are designed to have at least one reversible linkage in each chain capable of storing elastic energy.

The size of the *smallest relaxing molecular unit* within the network significantly affects the mechanical behavior of a CAN. The smallest relaxing molecular unit can be defined as either a reversible-*monomer* (Figure 3.2, left) or a reversible-*macromer* (Figure 3.2, right). Reversible-monomer CANs tend to have a uniform distribution of reversible crosslinks throughout the network topology. In contrast, reversible-macromer CANs contain long irreversible contiguous chains that are crosslinked by reversible covalent bonds. As shown in Figure 3.2, while the initial monomer composition is important, the polymerization mechanism has a major effect on whether the smallest relaxing molecular unit of a CAN is a monomer or a macromer.

The formation of a reversible-*monomer* network architecture typically follows a step-growth polymerization mechanism, which offers control over the mechanical properties by adjusting the initial monomer functionality (number of crosslinking functional groups per monomer). Moreover, the potential for enhanced concentration and uniform distribution of reversible crosslinks makes this strategy highly desirable for self-healing materials.

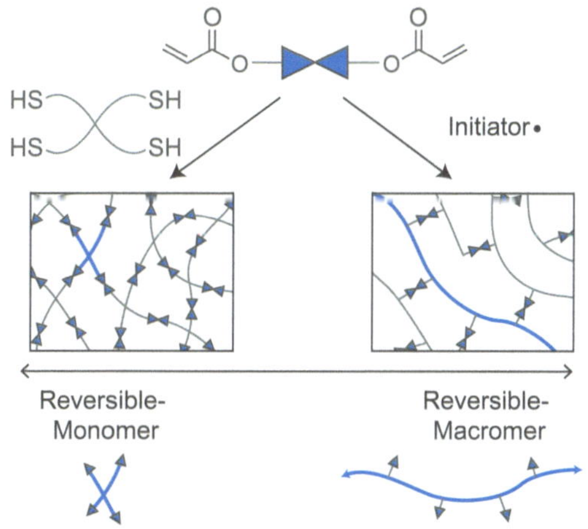

Figure 3.2 Two polymerization schemes from the same monomer possessing dynamic covalent functionality can produce CANs with two different smallest relaxing molecular units. A diacrylate monomer with dynamic covalent functionality can be copolymerized with a multifunctional thiol monomer in the presence of a base to produce a network possessing reversible-monomers (left) or homopolymerized in the presence of a radical to produce a network possessing reversible-macromers (right). Adapted from ref. 7 with permission from the Royal Society of Chemistry.

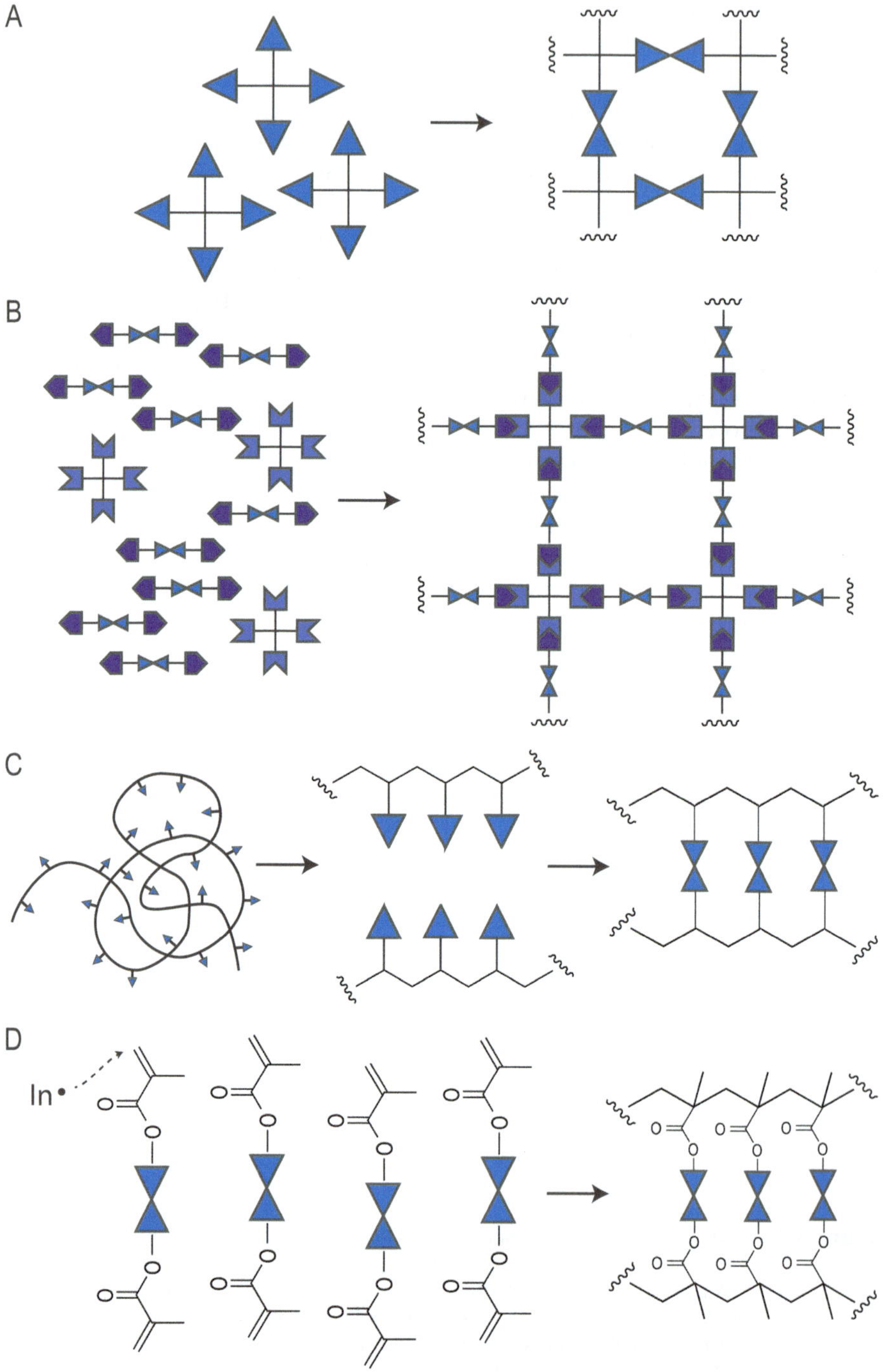
A
B
C
D
In•
O
O
O
O
O
O
O
O

The reversible covalent crosslinking reaction can either be part of the network forming chemistry (A) or latent within the backbone of the monomer (B). In the former case, there is at least one reversible linkage between every junction point or crosslink. Additionally, several suitable addition polymerization reactions can be used for crosslinking, such as disulfide formation (Section 3.4.2.1) and furan–maleimide cycloaddition adduct formation (Section 3.4.2.2). In the latter case, a wide range of reversible covalent linkages can be utilized so long as they are orthogonal (*i.e.*, do not interfere) with the polymerization reaction.

Reversible *macromer* network architectures are characterized by the presence of a long irreversible polymer chain capable of entanglement (*i.e.*, the macromer molecular weight exceeds the entanglement molecular weight), distinguishing them from high molecular weight 'monomers'. The presence of these long irreversible chains significantly influences the overall relaxation and adaptability of the polymer network. In the absence of crosslinks, the polymer would relax *via* reptation (*i.e.*, curvilinear diffusion). Therefore, the ultimate relaxation for macromer-based CANs is a balance between the timescale of reptation and the lifetime of the crosslink. While this complicates the relaxation process, the irreversible chains provide stability, while the reversible crosslinks allow the material to be remolded or repaired.

Reversible-macromer CANs are typically formed through two main approaches (Figure 3.3): by crosslinking pendant reversible functional groups (C) or by chain polymerization of multifunctional groups flanking a reversible functional group (D). In the former case, the polymer is synthesized with pendant functional groups capable of reversible crosslinking reactions. The latter case requires compatibility of the reversible linkage with the polymerizable functional group. Moreover, the polymerizable moiety must be difunctional, such as those produced by radical mediated polymerization of meth(acrylate) (as shown in D) or by ring-opening metathesis polymerization (ROMP) of norbornene.

3.3.2 Reversible Covalent Crosslink Type

The dynamic covalent crosslinks that constitute a CAN are classified by whether they follow an *associative* (Figure 3.1(B)) or *dissociative* (Figure 3.1(C)) mechanism, respectively. The terms associative and dissociative are most

Figure 3.3 Select covalent adaptable network forming strategies. The blue triangles and 'bowties' represent a reversible covalent crosslink; the purple irregular pentagon and forked square (B) represent non-reversible crosslinks; and the wavy line represents connections to the greater network. The network topologies depicted in panels A and B are reversible-monomer architectures, while those in panels C and D are reversible-macromer architectures. In panels A and C, monomers utilize reversible linkages to form the network, whereas panels B and D show reversible linkages in the macromer backbone, distinct from the polymerizable functional groups.

closely related to organometallic chemistry terminology, and it is now common practice to refer to CANs in these terms (*i.e.*, associative CAN or dissociative CAN). Incidentally, reactions that follow a concerted mechanism are often characterized as associative CANs, but they are more accurately described as interchange CANs. In Section 3.4, a selection of chemistries that constitute associative and dissociative reactions are examined, where crosslinks with an associative mechanism maintain network connectivity, while those with a dissociative mechanism disrupt it. In fact, the associative mechanism forms a higher network connectivity intermediate, although this distinction is of limited practical importance. The type of reversible crosslink mechanism leads to different dynamics on the molecular scale, resulting in distinct macroscopic mechanical behaviour (Section 3.5).

3.3.2.1 *Thermodynamic-* Versus *Kinetic-controlled Reversible Covalent Crosslinks*

Reversible crosslink chemistries are characterized as either thermodynamically or kinetically controlled. While the ultimate conversion of a reaction at equilibrium is temperature-dependent, the reaction rate may be insufficient to achieve equilibrium on practical timescales. Thermodynamically controlled covalent chemistry is often associated with dynamic covalent chemistry (DCC),[4–6] which is comprised of reactions at equilibrium and is susceptible to changes in thermodynamic variables, such as temperature, pressure, and concentration. Ideally, these reactions exhibit rapid kinetics, allowing them to spontaneously change to environmental changes. In contrast, kinetically controlled CANs are typically triggered by an event that initiates bond rearrangement. For example, polymer networks capable of radical-mediated bond exchange (*e.g.*, *via* addition–fragmentation chain transfer) remain inert until a radical is generated, typically by a photo-, thermo-, or redox initiator.

One critical difference between reversible covalent chemistry in small molecules and polymer networks is the phenomenon of vitrification. Below the material's glass transition temperature (T_g), the underlying molecular dynamics responsible for rearrangement become kinetically frozen until the temperature is elevated above the T_g. Therefore, when designing a material for a specific healing application, it is crucial to consider the material's state at a given temperature. Predictions of the T_g based on molecular parameters are not well developed; however, decreasing the polymer network flexibility and increasing side-group bulkiness generally increase the T_g. While vitrification can complicate certain healing strategies, it also provides an opportunity to control or trigger healing in a thermodynamically-controlled crosslinked polymer (*e.g.*, by elevating the temperature above the material's T_g). Materials that exploit glassy dynamics while possessing associative reversible covalent crosslinks above the T_g are often termed vitrimers[2] and exhibit 'strong' glass-forming behaviour akin to silica.[17,18] Nevertheless, as pointed out by Elling and Dichtel, dissociative CANs also exhibit similar thermal reprocessing behavior, rendering the vitrimer designation of little practical value.[19]

3.3.2.2 Triggering Reversible Crosslinking

The nature of dynamic covalent crosslinks, whether they are at equilibrium or triggered by an external stimulus, is a crucial consideration when tailoring materials for specific applications. CANs respond to mechanical stress or material failure through either *passive* or *active* mechanisms. In chemistries where reversible reactions are at thermodynamic equilibrium, the network undergoes continual rearrangement. This *passive* process can heal micro-fractures, blunt crack initiation sites, or simply reduce material defects that may have formed during fabrication. In contrast, *active* healing mechanisms are initiated in response to an external stimulus, such as light or mechanical stress. The latter stimulus (mechanical stress) is often referred to as *self-healing*, although all passive mechanisms also fall under this category.

One apparent limitation to passive thermoreversible CANs is the requirement to heat the material to accelerate the healing process. Typically, demonstrations of thermoreversible network healing involve inducing cracks in the sample or fracturing it followed by heating in an oven. The efficiency of healing in such materials heavily relies on heat transport, which is influenced by sample geometry and oven heating conditions. Moreover, these strategies are not easily applicable to many real-life applications, especially those involving large-scale fabrication or where material removal is challenging. In such cases, having a method to externally trigger the network rearrangement would be advantageous.

Light absorption can generate heat to initiate healing in a thermoreversible network.[20,21] Controlling heat production involves adjusting absorber concentration and light intensity. However, challenges arise from managing light penetration depth and heat transport complexities.

Another externally triggered heating option involves the internal heating of a CAN containing electrically conductive particles by applying an alternating electromagnetic field, inducing a rapid alternating current in these 'susceptors', resulting in Joule heating. This method, historically used in thermoplastics for welding, allows for greater penetration of the electromagnetic field into the material. Unfortunately, it faces similar challenges in heat transport as light-based methods.

A related externally triggered heating approach involves hysteresis heating of ferromagnetic nanoparticles. While Joule heating in nanoscale particles is typically insufficient to heat the material, the alternating electromagnetic field can induce magnetic domain switching within ferromagnetic materials resulting in significant heating. Ferromagnetic materials exhibit a temperature ceiling known as the Curie temperature, beyond which their magnetic domains randomize, causing the material to lose its magnetic character. By integrating ferromagnetic nanoparticles into a thermoreversible CAN, the material heats rapidly in the presence of an alternating electromagnetic field, but only up to the Curie temperature. This self-limiting heating mechanism prevents overheating and reduces the potential for thermally induced irreversible side reactions. This kind of strategy has successfully repaired a fractured thermoreversible Diels–Alder network embedded with chromium oxide nanoparticles multiple times without compromising mechanical strength.[22]

3.4 Reversible Covalent Crosslink Chemistries

Numerous reversible chemistries are suitable for the design of healable covalent networks. In particular, DCC[4–6] provides an excellent framework to guide the selection of reversible covalent crosslinks. Over the last decade, integrating DCC into polymer networks has opened new chemical pathways for creating new adaptable and healable crosslinked networks. The following sections highlight key reversible chemistries that are suitable for healable CANs. Attempting to list all such reactions is bound to lead to omissions, and in such a rapidly growing field, this kind of listing can quickly become outdated. Additionally, reversible chemistry is a mature field, making it challenging to attribute its development to any specific set of researchers without overlooking seminal studies from the early 20th century. Nevertheless, there are several reviews that outline DCC,[4–6,23] which can be implemented in CANs, or directly discuss the use of reversible chemistry in covalent networks.[1,3–12]

Reversible reactions can be categorized as *reversible addition* and *degenerative exchange* (*i.e.*, thermodynamically neutral) reactions (Figure 3.4). Reversible addition reactions (or reversible combination reactions) have an $A + B \leftrightarrow AB$ reaction structure, where A can be the same as B (*e.g.*, thiol in the reversible disulfide reaction, A) or different from B (*e.g.*, diene and dienophile in the reversible Diels–Alder reaction, B). These reactions necessarily result in a dissociative network strand rearrangement.

Degenerative exchange reactions[24,25] (or degenerative displacement reactions) follow an $AB + B' \leftrightarrow AB' + B$ reaction structure, where the prime indicates a different B group with the same chemical identity (Figure 3.4). Metathesis reactions, with the reaction structure $AB + A'B' \leftrightarrow AB' + A'B$, also fall under degenerative exchange reactions. Importantly, degenerative exchange reactions feature identical reactants and products, resulting in no thermodynamic driving force for either the forward or reverse direction.

Crosslinks based on degenerative exchange reactions induce associative network strand rearrangement (Figure 3.4). Interestingly, if the reversible crosslink follows a unimolecular nucleophilic substitution (S_N1) mechanism, the network technically undergoes dissociative strand rearrangement (*i.e.*, the transition state for an S_N1 reaction is uncrosslinked). However, the transition state is typically short-lived and has a negligible population at any given time. As a result, the mechanical properties (Section 3.5) more closely resemble those of associative CANs. Additionally,

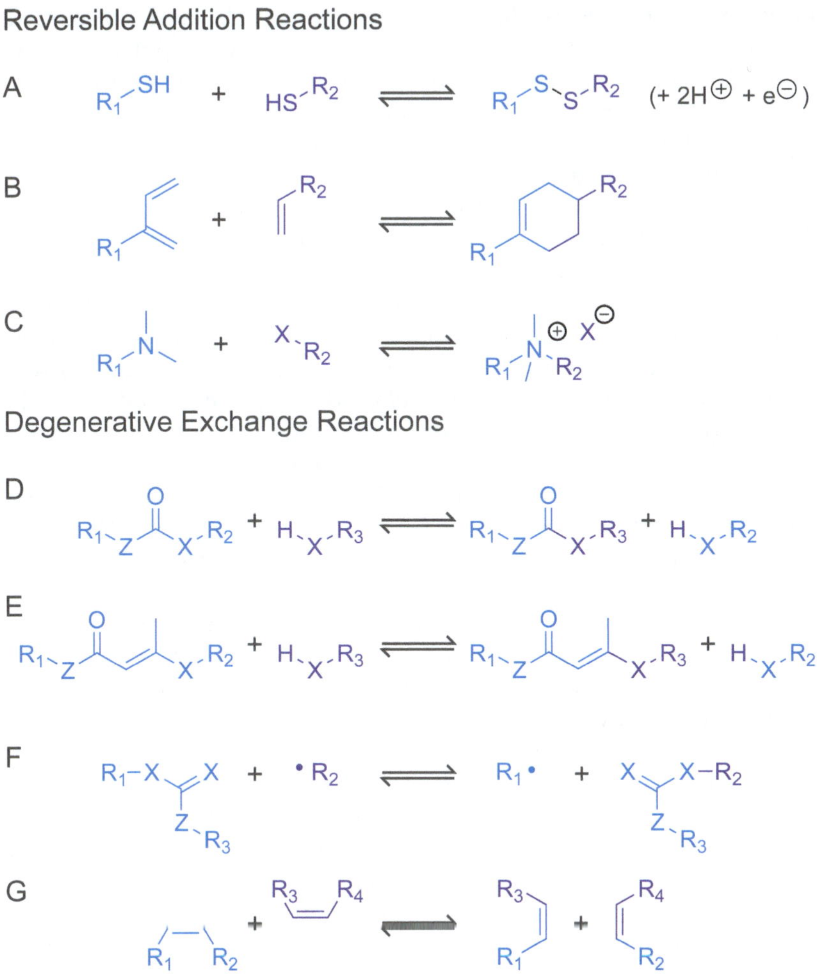

Figure 3.4 A selection of dynamic covalent chemistries for reversible crosslinks in covalent adaptable networks: (A) disulfide formation, (B) the Diels–Alder cycloaddition, (C) transalkylation (X = halide), (D) carbonyl-based *trans*-X reactions (*e.g.*, transesterification, Z=CH_2, X=O; transcarbamoylation, Z=NH, X=O; transcarbonation, Z=O, X=O; transthioesterification, Z=CH_2, X=S), (E) vinylogous carbonyl *trans*-X reactions (*e.g.*, vinylogous urethane, Z=O, X=NH; vinylogous urea, Z=NH, X=NH; vinylogous amide Z=CH_2, X=NH; vinylogous carbamothiolate Z=S, X=NH; keto-enamine, Z=CH_2, X=NH; dithioalkylidene Meldrum's acid, Z=O, X=S), (F) reversible addition–fragmentation, and (G) metathesis.

degenerative exchange-based crosslinks that follow a concerted mechanism, such as olefin metathesis (Section 3.4.3.4), also exhibit properties associated with an associative network (Figure 3.4).

Selection of the reversible crosslinking reaction type is one of the overarching considerations in the design of a CAN for a particular healing application. The reversible disulfide bond (A and discussed in Section 3.4.2.1) is one the first dynamic bonds explored in polymer networks, and it is also one of the more chemically complex crosslinks. In addition to reversible addition reactions, the disulfide is capable of degenerative exchange though several mechanisms, including metathesis. The reversibility and specificity of the Diels–Alder cycloaddition (B and discussed in Section 3.4.2.2) makes it one of the premier reversible addition reactions.[26,27]

The reactions shown in C, D and E are often listed as '*trans*-X' reactions, which encompass several dynamic covalent crosslinks, including those based on *trans-N*-alkylation (Section 3.4.2.3), transesterification (Section 3.4.3.1), transcarbamoylation,[28,29] transcarbonation,[30] and transthioesterification.[31,32] With the exception of the transalkylation reaction (C),[33–36] these *trans*-X reactions types are classified as degenerative exchange reactions. Vinylogous carbonyl *trans*-X reactions (Section 3.4.3.2) similarly exhibit degenerative exchange with heteroatom functional groups (*e.g.*, vinylogous urethane,[37,38] vinylogous amide,[37] vinylogous carbamothioate,[39] diketoenamine,[40,41] and dithioalkylidene Meldrum's acid[42–44]) and often exhibit fast and tunable bond exchange dynamics without the need for a catalyst.

The last two reaction examples are radical-mediated addition–fragmentation reactions (F and Section 3.4.3.3) and metathesis reactions (*G* and 3.4.3.4). Addition–fragmentation reactions are often overlooked as a mechanism for strand rearrangement, but they have played an important role in the renaissance of covalent adaptable networks[45] as well as providing a critical mechanism for radical triggered network healing. In addition to alkene metathesis, shown in *G*, there are numerous other reactions that are related to functional groups within the *trans*-X reaction category, such as ester–ester metathesis; however, one should note that several of the presumed metathesis reactions end-up following a reversible addition mechanism. For example, latent water in the network can facilitate ester hydrolysis, which can lead to a hydroxyl product that can undergo transesterification; thus, network strand rearrangement becomes dependent on the presence of water.

Only a small selection of reversible addition and degenerative exchange crosslinks is presented herein; these examples are given to provide insights into new CAN designs rather than putting limitations on chemical crosslink choices. In all cases, one must consider the potential limitations by detrimental irreversible side-reactions (discussed in Section 3.4.1) or low efficiency, the extent of which is often application dependent. Some other particularly noteworthy examples of reversible linkages that were omitted are acetal-based crosslinks (*e.g.*, *O*,*O*-acetals,[46] *S*,*O*-acetals,[47,48] *N*,*O*-acetals,[49] and *S*,*S*-acetals[44,50,51]), *trans*-'siloxanification'-based crosslinks,[52–55] imine crosslinks,[56–58] isocyanate–imidazole crosslinks,[59] thiol–Michael adduct crosslinks,[60,61] nitroxide-based crosslinks,[62–65] boron–oxygen crosslinks (*e.g.*, boronic esters,[66] dioxaborolanes,[67] boroxine,[68] *etc.*), and [2+2] and [4+4] cycloaddition adduct crosslinks (*e.g.*, coumarins,[64,69–72] cinnamates,[73–77] and thymines,[78] and anthracenes[75,79–81]). Moreover, metal–organic healing materials are an important part of the materials-healing paradigm but are not covered here. Nevertheless, as healing applications change, novel material implementations arise, or new reaction discoveries are made, any of these reaction types may be made practical as a reversible crosslink in covalent networks.

3.4.1 An Aside on the Reversibility of Dynamic Crosslinks

There are three main reaction characteristics to consider for dynamic crosslinks within a CAN for a particular application: type, rate, and side reactions. The following sub-sections primarily discuss reaction type, including whether the reaction leads to associative or dissociative strand rearrangement. Controlling the rate of strand rearrangement for these reaction types has been the focus of much recent literature. Efforts have centered on increasing the rate by incorporating functionality adjacent to the dynamic bond or by introducing a catalyst. For the latter, it is important to consider the specific application, where the environmental conditions could lead to catalyst deactivation or catalyst leaching over time.

The loss of network adaptability due to irreversible side reactions is a critical obstacle in healing applications, limiting the number of effective healing cycles and decreasing the lifetime of the material. Some applications may require conditions that are not suitable for prolonged reversibility of certain reactions. For example, Diels–Alder-based networks that are crosslinked by furan and maleimide are often seen as exemplary dissociative CANs; however, the adduct undergoes irreversible aromatization under prolonged exposure to high temperatures.

Healable CANs typically exclude reversible condensation crosslinking reactions, since the production of a condensate is a leachable component that is not connected to the network.[1,3] For example, the condensation reaction between multifunctional aldehyde and amine monomers produces a network crosslinked *via* an imine functional group (Schiff base) with water as a by-product (Figure 3.5(A)), making the imine crosslinks unstable in the presence of water. In the absence of water, the imine functional group will undergo bond exchange with pendent amine functional groups, leading to network adaptability and healing (Figure 3.5(B)). Therefore, the application and environmental context are critical aspects when selecting reversible covalent crosslinks for a healable CAN. In materials recovery and recycling, the ability to break down a network to its constituent monomers in the presence of condensate may even be considered a desirable feature.

3.4.2 Reversible Addition Reactions

Reversible addition reactions have an $A + B \leftrightarrow AB$ structure, where A and B can be the same or different functional groups. The general bond exchange scheme is illustrated in Figure 3.6, depicting a characteristic equilibrium

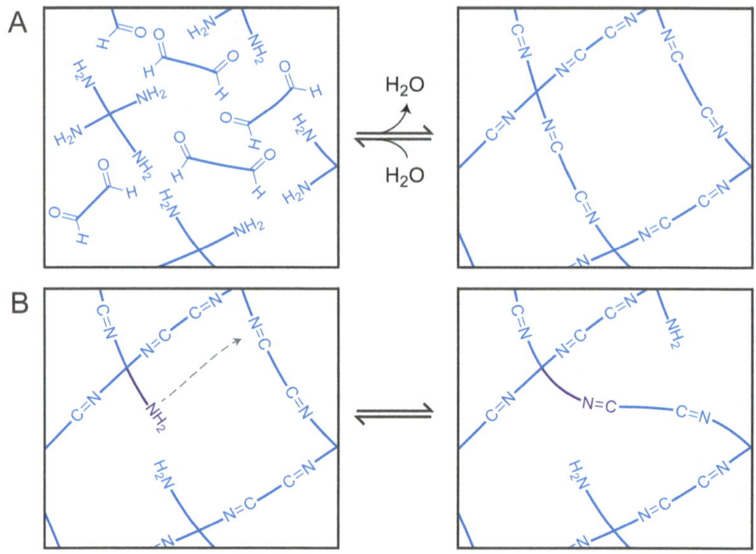

Figure 3.5 Formation (A) and exchange (B) reactions of an imine polymer network. The imine network is formed by the reaction of aldehyde and amine, requiring the removal of water. When the network is formulated with excess amine, an exchange reaction with imine functional groups in the network backbone provides a route to network strand rearrangement.

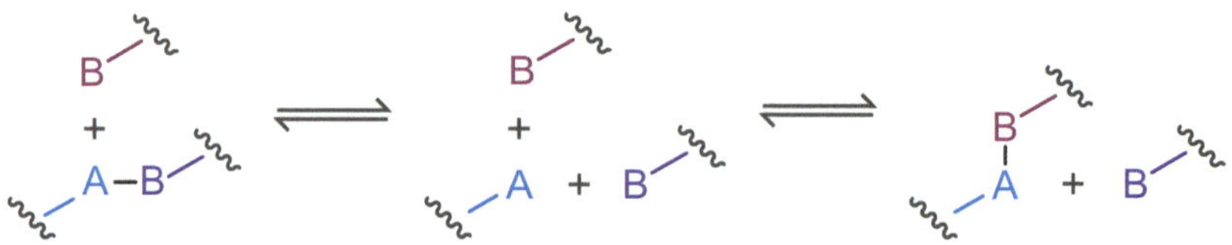

Figure 3.6 General bond exchange scheme for reversible addition crosslinks, where A and B are functional groups that can be the same (*e.g.*, thiol) or different (*e.g.*, diene and dienophile).

state that is uncrosslinked. If one were to focus solely on the far left and right equilibrium states shown in Figure 3.6, they might mistakenly classify this as a degenerative exchange reaction. However, unlike degenerative exchange, reversible addition reactions include an equilibrium with an uncrosslinked state. The proportion of this state varies with thermodynamic parameters such as temperature and functional group concentration. The thermodynamic influence on the crosslink density and the associated rubbery modulus distinguishes reversible addition reactions from degenerative exchange reactions. Three examples of reversible addition crosslinks are given below: disulfide crosslinks (Section 3.4.2.1), the Diels–Alder adduct crosslinks (Section 3.4.2.2), and ammonium/iminium cationic crosslinks (Section 3.4.2.3).

3.4.2.1 Disulfides

Reversible disulfide formation in covalent networks is the prototypical CAN crosslinking reaction. Sulfur vulcanization is one of the earliest commercial successes in polymer networks, and the reversibility of sulfur-crosslinked materials has been extensively studied. As mentioned above (Section 3.2), the general concept of CANs is rooted in the seminal work of Tobolsky and co-workers, who examined the phenomenon of stress relaxation in disulfide (and tetrasulfide) crosslinked materials.[14,82,83] Historically, reversible crosslinking was largely considered to be a negative attribute. Today, this same attribute in disulfide crosslinked networks is being exploited in self-healing applications.[84–87] While the dynamic nature of the disulfide bond has been more broadly implemented in small molecule or functional tether applications (*e.g.*, the release of bioactive molecules in drug delivery[88–91] or in dynamic combinatorial libraries[92]), such concepts are readily extended to impart a range of responsiveness into the network *via* reversible disulfide crosslinks.

The disulfide bond is an extremely versatile reversible linkage, capable of undergoing homolytic breaking and reforming as well as exchange reaction with ether thiolate anions[93–95] or thiyl radicals.[82,86] The formation of the thiolate from a thiol is dependent upon its pK_a, and thus the network rearrangement is controlled *via* the pH of the surrounding media (a pH responsive material). The thiyl radical is typically generated *via* the radical abstraction of the hydrogen, which is facilitated by a photo-, thermo- or redox-initiator. Moreover, the formation of disulfide crosslinked polymers is readily achieved by the oxidation of thiol monomers or macromers (*e.g.*, using I_2 or $FeCl_3$) and can also be reduced to their constituent parts (Figure 3.7(A)).[96]

While disulfide crosslinked polymers have been utilized for over a century, the deliberate incorporation of disulfide crosslinks for reversible network formation is a relatively recent endeavor. Unlike the uncontrolled incorporation of sulfur *via* vulcanization, polymer chemists have systematically incorporated disulfide crosslinkers. Chujo *et al.*[97] demonstrated the reversible oxidation and reduction of disulfide crosslinked poly(*N*-acetylethylenimine) using a crosslinker with the disulfide in the backbone. Subsequently, Canadell *et al.*[87] showed healing in a disulfide network fabricated with a multifunctional thiol and difunctional epoxy containing a disulfide within the monomer backbone (*e.g.*, Figure 3.7). They utilized the ability of the disulfide linkage to undergo

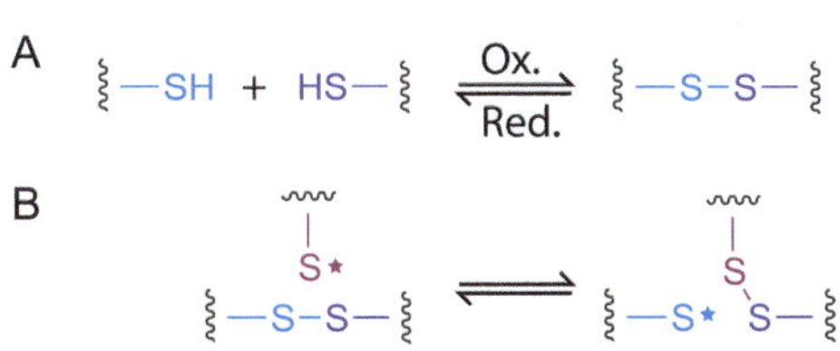

Figure 3.7 (A) Reversible addition disulfide formation from thiol functional groups and (B) subsequent regenerative exchange reactions, where the star represents either a radical or anion.

disulfide–disulfide exchange reactions, where bond rearrangement involves breaking a disulfide bond and the subsequent thiolate reaction with another disulfide bond (Figure 3.7(B)). Others have demonstrated that direct disulfide–disulfide metathesis can be achieved in small molecules using a phosphine catalyst,[98,99] which is readily extended to disulfide-based CANs. Furthermore, Odriozola and co-workers[100,101] demonstrated stress relaxation in elastomeric networks utilizing the metathesis of aromatic disulfide linkages.

Photo-mediated connectivity rearrangement of a disulfide crosslinked PEG hydrogel was demonstrated by Fairbanks *et al.*[86] Here, the network was directly formed by the oxidation of thiols on a 4-armed PEG in water using a hydrogel peroxide/sodium iodine solution. A photoinitiator was diffused into the hydrogel and the material was subsequently strained and photopatterned (*i.e.*, mechanophotopatterning). Utilizing this technique, the material's equilibrium geometry was manipulated using masked irradiation. Additionally, when two hydrogel specimens were pressed together and irradiated, the radical-induced rearrangement of disulfide bonds resulted in the fusion of the pieces into a single hydrogel piece, demonstrating the material's healing potential. It is noteworthy that the observed stress relaxation cannot simply be explained by photolytic cleavage of the disulfide bond (*i.e.*, dissociative mechanism); rather, the relaxation follows an associative mechanism related to thiyl radical addition across a disulfide (Figure 3.7(B)), similar to addition–fragmentation of the RAFT-based materials (reaction F in Figure 3.4 and discussed below in Section 3.4.3.3).

3.4.2.2 The Diels–Alder Reaction

Among the various cycloaddition crosslinking reactions, the thermoreversible [4+2] cycloaddition reaction, or Diels–Alder (DA) reaction, stands out for its robustness and easy implementation, even achieving 'click' reaction status.[26] The DA reaction is reasonably tolerant of a range of environmental conditions, such as water and oxygen, and its reactivity is readily tuned by the appropriate selection of conjugated diene and electronically activated double bond or dienophile.[102–104]

The initial applications of the Diels–Alder reaction within a covalent adaptable network dates back to the 1960s when Craven crosslinked a polymer having pendant furan functional groups with a multifunctional maleimide monomer (Figure 3.8(A)).[15] This thermoreversible crosslinked polymer can be post-formed upon heating the material to 120 to 140 °C. Since then, numerous researchers have used this thermoreversible crosslinking strategy to produce reversible macromer network-architectures.[105–114] It should be noted that this material does not undergo reverse gelation at these temperatures, but is nevertheless remoldable (or healable) while roughly maintaining its structure. Typically, the reversible gelation temperature for these reversible macromers is impractically high, triggering irreversible side reactions that limit its healing ability.

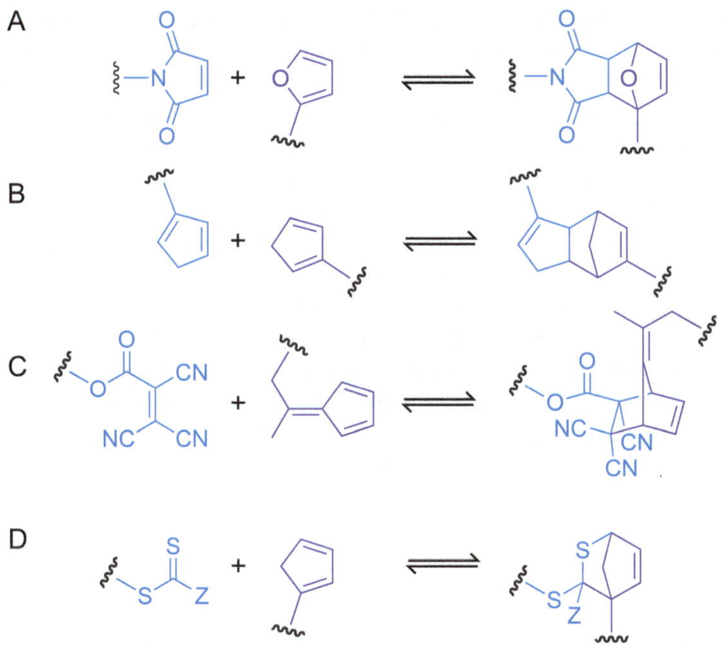

Figure 3.8 Select Diels–Alder crosslinks: (A) maleimide–furan, (B) cyclopentadiene dimerization, (C) tricyanoacrylate–fulvene, and (D) RAFT agent–cyclopentadiene.

There have been several implementations of maleimide–furan dynamic linkages in polymers.[27] Utilizing tetrafunctional furan and trifunctional maleimide monomers, Chen *et al.*[115] demonstrated the healing capability of thermoreversible CANs using a reversible monomer network formation strategy. In their study, the material was subjected to fracture toughness testing and then healed by heating the specimen for a prescribed amount of time (*e.g.*, 2 hours at 120 °C). During the thermal treatment, the material transformed from a glassy to rubbery polymer matrix ($T_g \sim 100$ °C),[116] enabling the molecular mobility of the functional groups and rearrangement of the DA crosslinks. At 120 °C, the sample maintained its shape and could be mended, exhibiting a healing efficiency of up to 50%. At sufficiently elevated temperatures, this material would, in principle, undergo a gel-to-sol transition; however, irreversible side reactions at these temperatures often limit their healing capacity over time. Nevertheless, the gel-point conversion (*i.e.*, the gel-to-sol transition) can be tailored by adjusting the number of functional groups in accordance with the Flory–Stockmayer equation.[117]

Furan and maleimide exhibit excellent kinetic and thermodynamic properties for a wide range of healing materials' applications, where these properties can be tuned through the selection of diene and dienophile pairs. Cyclopentadiene, for instance, can act as both a diene and dienophile, enabling the crosslinking of linear polymers *via* pendant cyclopentadienyl group dimerization (Figure 3.8(B)).[118] Furthermore, when cyclopentadiene is conjugated with an ester, the dicyclopentadiene dimer can react with another cyclopentadiene to form a trimer, enabling crosslinking of difunctional cyclopentadiene monomers.[119] Lehn and co-workers have explored fulvene and cyanoethylenecarboxylate (Figure 3.8(C)) as a diene-dienophile pair, which has rapid kinetics at room temperature.[120,121] They demonstrated rapid self-healing (~10 seconds) by simply pressing two pieces of this material together.[121] Moreover, the material did not 'stick' to other materials, highlighting the orthogonality of the DA rearrangement. As a final example, Barner-Kowollick and co-workers have investigated the hetero-DA reaction of a thio-thiocarbonyl 'RAFT' agent with cyclopentadiene (Figure 3.8(D)), which exhibited faster kinetics than the maleimide–furan cycloaddition.[122] The utilization of a difunctional RAFT agent enables the controlled synthesis of the polymer network backbone along with facile reversible crosslinking using a multifunctional cyclopentadiene monomer.

3.4.2.3 *Trans*-N-*alkylation and Ionenes*

Ionenes are polymers that are formed from multifunctional halide and amine monomers *via* the Menschutkin reaction, resulting in quaternary-ammonium crosslinks (Figure 3.9).[35,123,124] The incorporation of ammonium groups creates charged polymers, which are suitable for use in ion exchange applications, such as fuel cell membranes. The thermoreversible nature of these materials enables depolymerization at elevated temperatures, leading to dissociative relaxation behviour.[125] Ruckenstein *et al.*[34] have utilized both difunctional halide and amine monomers to crosslink polymers with pendant amine and halide functional groups, respectively. While the reversible kinetics depended on the specific amine and halide used, most of the gelled materials were able to 'flow' at elevated temperatures (215 °C) despite the presence of side reactions, such as Hoffman degradation.[34]

Recently, a range of exchangeable charged linkages, such as anilinium,[126,127] pyridinium,[128,129] or sulfonium salts,[130–132] have been considered in polymer networks. Using a variation on the ionene strategy, Obadia *et al.*[33,133] created a crosslinked triazolium CAN (Figure 3.10(A)). In their initial work, polymers were formed using an α-azide-ω-alkyne monomer through the copper(I)-catalyzed azide–alkyne cycloaddition (CuAAC) reaction and were subsequently crosslinked with 1,6-dibromohexane.[33] Despite being a so-called '*trans*-X' reaction (Section 3.4.3.1), this *trans*-*N*-alkylation

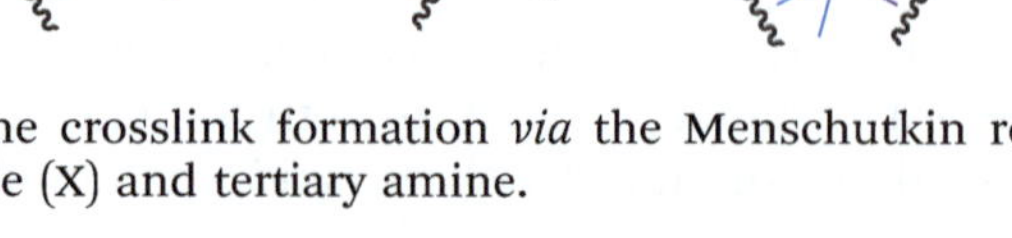

Figure 3.9 Ionene crosslink formation *via* the Menschutkin reaction between a halide (X) and tertiary amine.

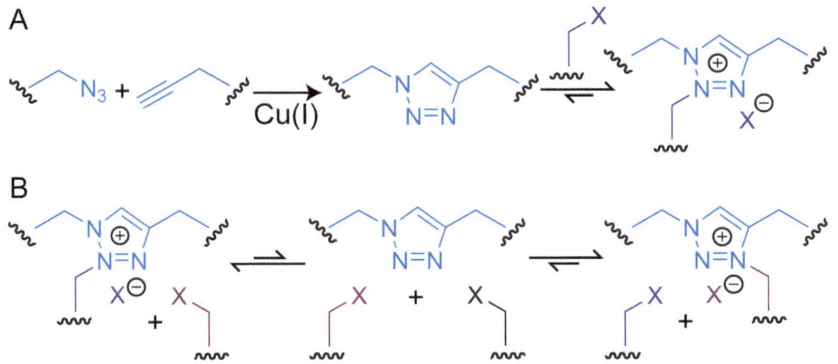

Figure 3.10 (A) Polymer formation *via* the copper(I)-catalyzed azide–alkyne cycloaddition (CuAAC) reaction followed by alkyl halide crosslinking, where X is a halide. (B) The resultant triazolium is in equilibrium with its crosslinked triazole form, enabling bond exchange with other nearby alkyl halides.

mechanism results in a dissociative mechanism for network strand rearrangement, as implied in Figure 3.10(B).[36] Some of the initial triazolium networks exhibited a relatively minor crosslink dependence on temperature and were presumed to follow an associative mechanism; however, this was later attributed to the modest temperature dependence of the equilibrium constant over the temperature range examined.[33,36,133] This illustrates how the differences between networks crosslinked with reversible addition reactions and degenerative exchange reactions can exhibit very similar behavior within finite temperature regimes.

3.4.3 Degenerative Exchange Reactions

Degenerative exchange reactions are displacement reactions where the products and the reactants are identical and thus thermodynamically neutral (Figure 3.11). While the kinetics of degenerative exchange reactions are affected by reaction conditions, such as concentration, temperature, and the presence of a catalyst, the state of crosslinking remains unchanged. Here, four examples of degenerative exchange reactions will be highlighted: transesterification reactions (Section 3.4.3.1), vinylogous carbonyl-based exchange reactions (Section 3.4.3.2), radical-based addition–fragmentation chain transfer reactions (Section 3.4.3.3), and olefin metathesis reactions (Section 3.4.3.4).

3.4.3.1 Transesterification Reactions

Transesterification reactions have been utilized in polymer science for several decades and have been extensively studied as a mechanism for relaxation in polymer blends.[134–137] While the transesterification reaction is

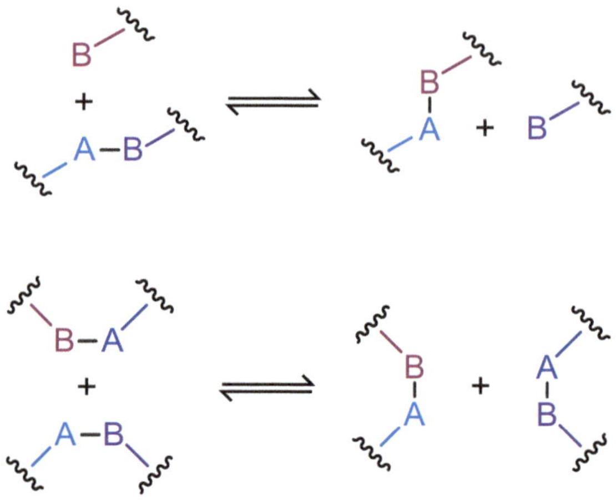

Figure 3.11 Single displacement (top) and double displacement, *i.e.*, metathesis (bottom) degenerative exchange reactions, where A and B are functional groups.

typically written between an ester and an alcohol, exchange is also possible between carboxylic acids or other esters. The catalyzed transesterification reaction in covalently crosslinked networks has been instrumental in bringing attention to the field of CANs, showcasing a facile route for rearrangement and self-healing. Specifically, Montarnal *et al.*[17] have demonstrated an epoxy-based covalent network that exhibits significant bond rearrangement *via* transesterification, leading to remolding and healing in what is typically an intractable material. Utilizing a crosslinking reaction between multifunctional epoxy and carboxylic acid monomers, the resultant network possessed both an ester functional group in the polymer backbone and a pendant hydroxyl functional group (Figure 3.12(A)). Each network strand has the potential to undergo a transesterification reaction with a neighboring strand (Figure 3.12(B)).

The specific transesterification example given in Figure 3.12(B) was selected both for its historical significance and to illustrate the complexity arising from a mixture of functional groups. There is a subtle chemical difference between the reactants and products in Figure 3.12(B): only the equilibrium state on the right contains a primary hydroxyl functional group. Additionally, there is a topological difference between the two states: one state has two difunctional stands, while the other has a trifunctional strand and a dangling end. Given sufficient time, the network will reach an equilibrium topology determined by the differing reactivity of the primary and secondary hydroxyl groups. However, the equilibrium state of ester crosslinks involving primary hydroxyl groups *versus* those involving secondary hydroxyl groups can exhibit slightly different dependencies on reaction conditions, such as temperature. While transesterification is indeed a regenerative exchange reaction, the presence of two reactive groups can result

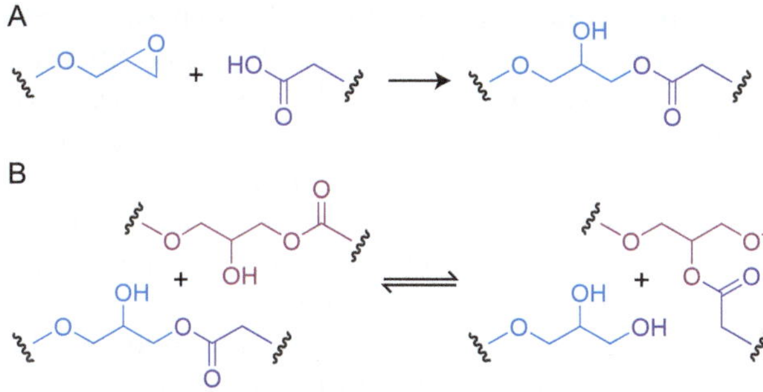

Figure 3.12 (A) Epoxy-carboxylic acid network formation and (B) rearrangement reactions. The formation of the polymer network between multifunctional epoxy and carboxylic acid monomers results in ester and hydroxyl bond formation. The hydroxyl and ester functional groups can undergo transesterification reactions resulting in network strand rearrangement.

in different topological states as a function of temperature. Therefore, in such complex chemical systems, interpreting the underlying bond exchange mechanism (*i.e.*, dissociative *versus* associative) based on the temperature dependence of mechanical properties, which is influenced by topology, could be misleading.

The relaxation behavior of the CAN above its T_g is directly influenced by the rate of bond exchange, emphasizing the pivotal role of catalyst selection in tuning network strand rearrangement. Montarnal *et al.*[17] used a Lewis acid catalyst (zinc acetate) to accelerate the kinetics by three orders of magnitude at 100 °C compared to ambient conditions, facilitating thermoplastic-like processing at elevated temperatures and thermoset-like stability at ambient conditions. Zinc serves multiple catalytic roles: enhancing nucleophilicity of the alcohol, increasing electrophilicity of the carbonyl, and coordinating the reactants for an efficient reaction.[138] Subsequently, Brutman *et al.*[139] demonstrated that another Lewis acid, stannous(II) octoate, further enhanced the reaction kinetics. Additionally, Self *et al.*[140] showed that Brønsted acids were also effective catalysts for transesterification in CANs.

Unsurprisingly, composition and mobility significantly influence exchange rates. Altuna *et al.*[141] demonstrated uncatalyzed relaxation in networks with a large excess of hydroxyl functional groups, albeit at a rate five times slower than the zinc-catalyzed process by Montarnal *et al.*[17] Ultimately, the relaxation rate can be finely tuned through careful control of composition and catalysis, providing multiple avenues to precisely tailor network dynamics for specific application time-scales.

3.4.3.2 Vinylogous Carbonyl-based Exchange Reactions

Compounds containing an α,β-unsaturated carbonyl functional group are classified as vinylogous carbonyls. Vinylogous esters, which are excellent

Michael acceptors, are frequently employed in network formation with multifunctional nucleophiles such as thiols. The thiol–Michael adduct is well known to undergo reversible addition[142] and has been used in CAN applications.[60] When the vinylogous carbonyl is conjugated with an amine, forming a vinylogous urethane, it can undergo an uncatalyzed exchange reaction with other amine functional groups (*e.g.*, see Figure 3.13(A)), offering a simple approach for creating vinylogous-based networks suitable for healable and recyclable materials applications.

One drawback to creating vinylogous urethane networks with acetoacetate-based monomers is the production of water as a byproduct, necessitating a slow, low-temperature (<100 °C) curing step to minimize porosity and imperfections. Spiesschaert *et al.*[143] demonstrated that the same product could be achieved through an addition reaction using a 2-butynoate functional monomer (Figure 3.13(B)). Concurrently, Guo *et al.*[144] also showcased the use of propiolate functional monomers to produce a similar product through an amine-yne 'click' reaction. This less sterically hindered vinyl functional group resulted in a lower activation energy for exchange (Figure 3.13(C)).[144,145] It should also be noted that there are several other vinylogous carbonyl-based exchange reactions suitable for healing materials applications, such as vinylogous amides,[37] vinylogous carbamothioates,[39] and vinylogous ureas.[146]

Cyclic vinylogous carbonyls have been shown to have exceptional bond exchange kinetics and a reportedly low occurrence of side reactions. The diketoenamine reversible linkage is easily integrated into a network by reacting a multifunctional amine with a β-triketone (Figure 3.14(A)).[40,41] Similar to the preparation of vinylogous urethane-based networks using

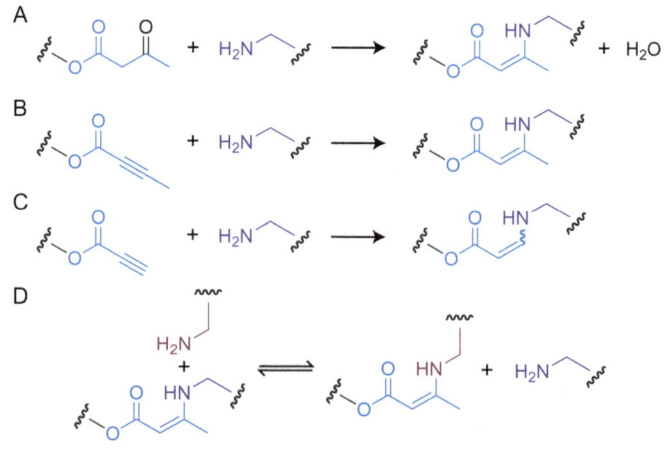

Figure 3.13 Three routes for vinylogous urethane formation: reaction between (A) acetoacetate and amine functional groups, (B) 2-butynoate and amine functional groups, and (C) propionate and amine functional groups. (D) Network strand rearrangement through vinylogous urethane exchange reaction with amine.

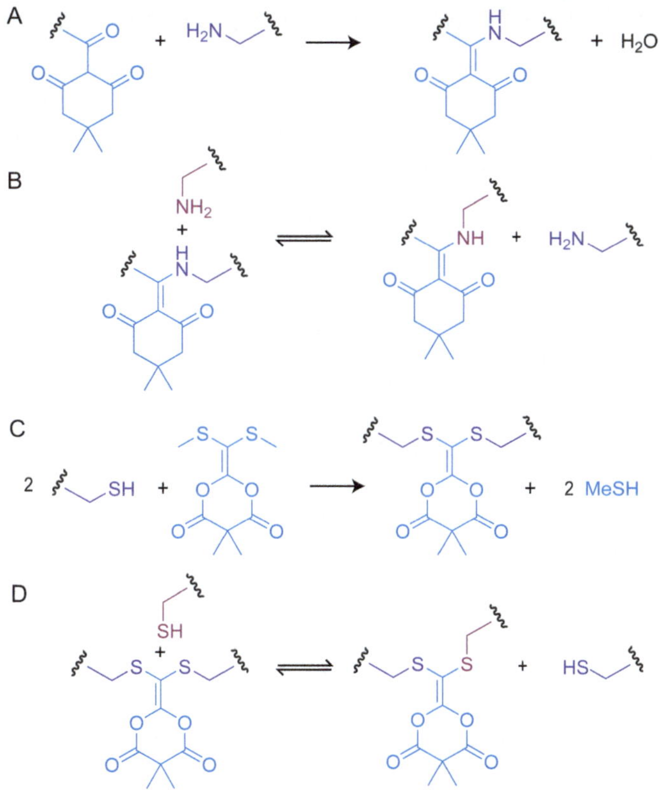

Figure 3.14 Two examples of cyclic vinylogous carbonyl compounds that undergo exchange are diketoenamines and Meldrum's acid-based dithioalkylidenes, the preparation of which is shown in A and C. Within the network strand, the diketoenamine can undergo exchange with a pendant amine functional group (B), whereas the Meldrum's acid-based derivative can undergo exchange with a pendant thiol functional group (D).

acetoacetate, network formation produces water as a condensate byproduct that must be removed; however, this reaction is highly efficient, with a click reaction-like quality. In the presence of excess amine, the exchange reaction proceeds rapidly without a catalyst (Figure 3.14(B)).

Another cyclic vinylogous carbonyl that has garnered recent attention is Meldrum's acid derivatives. Anslyn and co-workers[42] demonstrated that Meldrum's acid exhibits click-like reactivity and subsequent exchange with alkyl thiols. Subsequently, Kalow and co-workers showcased Meldrum's acid as a dynamic crosslink in CANs (Figure 3.14(D)).[43,44] Similar to diketoenamines, this reaction results in a condensate product (methanethiol, as shown in Figure 3.14(C)). El-Zaatari *et al.*[44] utilized various derivatives of Meldrum's acid-based dithioalkylidenes to crosslink polydimethylsiloxane, demonstrating the ability to tune the relaxation times over a wide range.

3.4.3.3 Radical-mediated Addition–Fragmentation Reactions

The implementation of radical-mediated addition–fragmentation chemistry[147] within polymer networks is a fairly recent development.[45] This is most likely due to the fact that reversible addition–fragmentation chain transfer (RAFT) is typically associated with the synthesis of low polydispersity linear polymers or block copolymers. Interestingly, the work of Scott *et al.*[45] is one of the first modern developments of covalent adaptable networks. One of the particularly attractive features of using addition–fragmentation is the on-demand nature of bond rearrangement that is achieved using light *via* a photo-radical generator (*i.e.*, a photoinitiator).

The concept of radical-mediated reversible addition–fragmentation was recognized several decades ago *via* the intramolecular homolytic substitution (S_H2') reaction of an allylic functional group.[148–151] The first utilization of this mechanism to control polymerization was in the 1980s;[152–154] however, the full potential of the RAFT strategy to control polymerization polydispersity was not realized until the late 1990s.[155] An essential aspect of the RAFT strategy is that the reactants (before addition) and the products (after fragmentation) are identical, except for possibly the molecular weight (*i.e.*, RAFT is a degenerate chain transfer process; see F); therefore, there is no thermodynamic direction to the reaction coordinate (*i.e.*, without termination, the reaction would continue to exchange). It should be noted that non-degenerative chain transfer processes may also be an effective strategy for network rearrangement, but radical generation and termination ultimately limit the extent of chain transfer events.

The first demonstration of addition–fragmentation chain transfer in a covalently crosslinked network utilized an allyl sulfide functional group within the network strands.[45] In this study, Scott *et al.* incorporated an allyl sulfide ring-opening monomer into a thiol–ene network formed by photopolymerization (Figure 3.15(A)). The allyl sulfide ring-opening monomer, previously used to form linear polymers,[156–158] acts as a chain extender within a thiol–ene step-growth polymerization (Figure 3.15(B)). Utilizing a latent photoinitiator left over from the network formation step, they triggered addition–fragmentation within the polymer network backbone, yielding photoinduced plasticity. Importantly, they demonstrated that the material properties, specifically the elastic modulus, were statistically identical before and after the network rearrangement, showing that the photoplasticity was not due to photodegradation.

The allyl sulfide functional group is well-suited for radical-mediated thiol–ene polymerization due to the degenerate chain transfer mechanism with thiyl radicals (Figure 3.15(C)).[152,159] In an ideal thiol–ene polymerization, the carbon radical intermediate does not react with other vinyl functional groups (*i.e.*, no homopolymerization) but rather abstracts a hydrogen from the thiol (Figure 3.15(A)). Exploiting this concept, Bowman and co-workers demonstrated that such a network rearrangement strategy has the additional benefit of dramatically reducing polymerization-induced shrinkage stress by

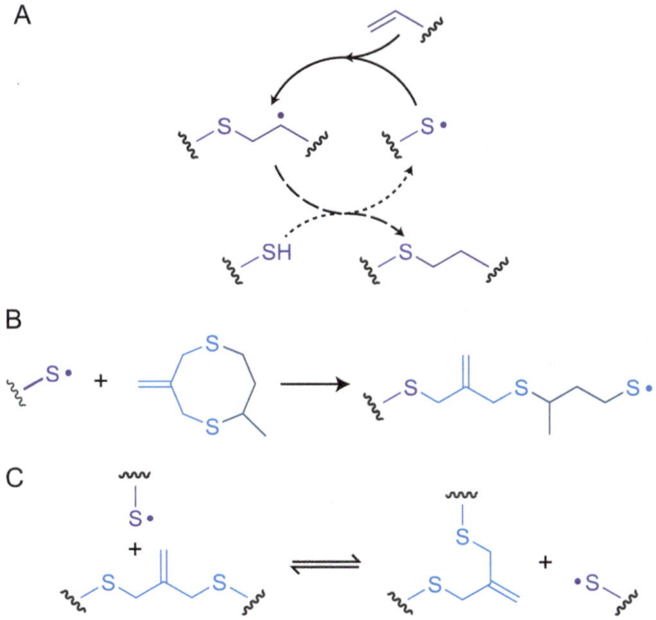

Figure 3.15 Thiol–ene reaction mechanism (A), allyl sulfide insertion through chain extension (B), and strand rearrangement through allyl sulfide addition–fragmentation (C). Network formation using the thiol–ene mechanism (*i.e.*, using multifunctional thiol and ene (vinyl) monomers) results in a thiyl radical intermediate. (B) In the presence of 2-methyl-7-methylene-1,5-dithiacyclooctane (MDTO), an allyl sulfide is readily incorporated within the network without altering the monomer stoichiometry. (C) The thiyl radical reaction with allyl sulfide (addition–fragmentation) preserves the overall chemistry (*i.e.*, degenerative exchange) but results in a change in bond connectivity.

utilizing linear monomers with an allyl sulfide backbone.[159] They further demonstrated that a non-degenerative scheme (*i.e.*, the carbon radical addition to the allyl sulfide) can still significantly reduce the polymerization stress, even in materials that exhibit glassy dynamics.[160–163] Finally, they demonstrated other applications for polymer networks containing allyl sulfide, such as mechanophotopatterning, which enables photo-induced surface topology and shape changes to a covalent network after network fabrication.[164,165] It should be noted that all the crosslinked polymers described herein not only have the capacity to heal but also to undergo a change in the materials equilibrium shape.

The addition–fragmentation chain transfer concept was further expounded upon using a conventional RAFT chain transfer agent (CTA) (Figure 3.16). Matyjaszewski and co-workers utilized a trithiocarbonate (TTC) functional group in the backbone of a di(methylmethacrylate)-based monomer to produce a polymer network having RAFT functional groups in the crosslink *via* a chain-growth mechanism (*i.e.*, a reversible macromer architecture).[166,167]

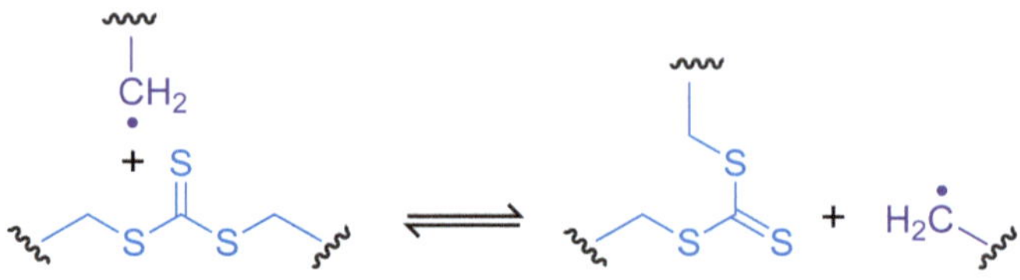

Figure 3.16 Degenerative transfer reaction between a carbon radical and a trithiocarbonate.

The TTC functional group has been used as a RAFT agent in a number of linear polymer fabrication strategies[168–172] and is capable of degenerative chain transfer with carbon-centered radicals. In the first demonstration of healing using this approach, a polymer gel was formulated using methyl methacrylate and TTC dimethacrylate monomers in the presence of AIBN and swollen in anisole. Deoxygentated styrene and CuBr/pentamethyldiethylenetriamine were introduced to three discrete pieces of this fabricated gel and heated to 60 °C for four hours. A styryl radical (*i.e.*, benzylic radial) was added across the TTC functional group, mending the discrete pieces into a single crosslinked gel.

Amamoto *et al.* have further demonstrated the reshuffling of the TTC functional group by simply irradiating the material with UV light.[166,173] The inclusion of a photo-radical generator is unnecessary to trigger the photo-induced network rearrangement, as the TTC functional group is capable of direct photo-cleavage of the relatively weak carbon–sulfur bond.[174] In principle, this strategy enables material adaptation to be photo-triggered an unlimited number of times (*i.e.*, assuming no side reactions). Nevertheless, the wavelength and intensity necessary to generate a sufficient number of radicals to trigger the addition–fragmentation cascade are limiting and likely causes other irreversible reactions. To address this, Amamoto *et al.*[175] utilized a thiuram disulfide crosslinked material, which is readily cleaved using wavelengths into the visible spectrum. Upon irradiation, the relatively weak thiuram disulfide bond is cleaved, producing two stable (*i.e.*, long-lived) thiuryl radicals (Figure 3.17(A)). While this process appears to follow a dissociative mechanism, Amamoto *et al.* argue that the thiuryl radical is able to undergo subsequent addition-fragmentation with thiuram disulfide functional groups, thereby undergoing a cascading bond-exchange, associative mechanism (Figure 3.17(B)). Moreover, the network was constructed using isocyanate-alcohol chemistry, producing a network through a step-growth mechanism with a reversible monomer architecture. The authors utilized this attribute to control the number of TTC functional groups per crosslink, maximizing the addition–fragmentation network rearrangement effect.

The utilization of addition–fragmentation chain transfer in covalent adaptable networks is still relatively unexplored. Beyond the allyl sulfide and TTC functional groups, there are a large number of RAFT chain transfer agents that are suitable for implementation into the backbone of the

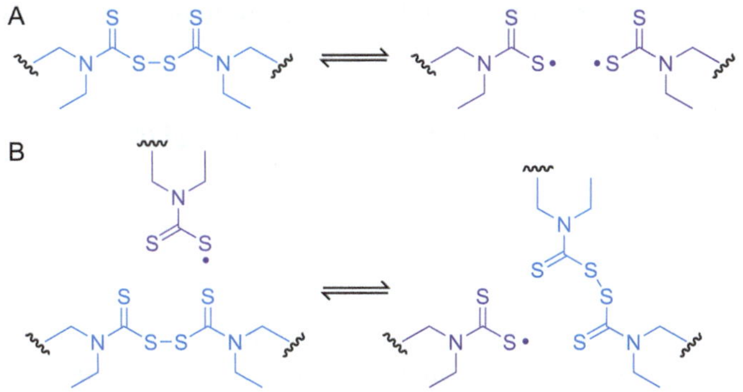

Figure 3.17 (A) Formation of thiuryl radicals *via* homolytic cleavage and (B) subsequent addition–fragmentation of a thiuriam disulfide functional group.

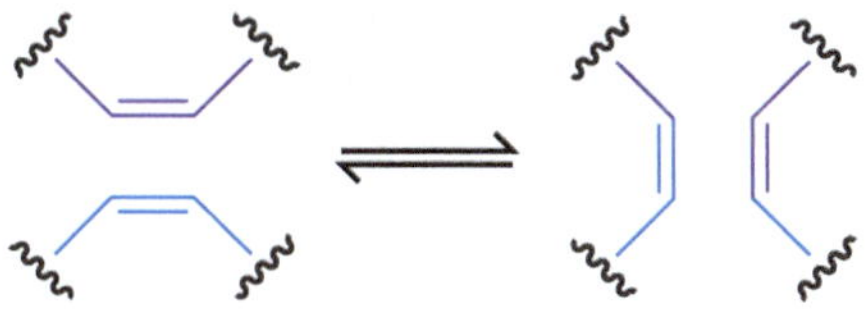

Figure 3.18 Olefin metathesis reaction.

polymer network.[176–179] There is no requirement for the addition–fragmentation crosslink to be symmetrical. In general, the Z and R groups of the base RAFT chain transfer agent can be appended with polymerizable functional groups (*e.g.*, F). Moreover, the selection of the Z functional group allows precise control over the reactivity of the addition–fragmentation crosslink.

3.4.3.4 *Olefin Metathesis*

Olefin metathesis has been extensively used in small molecule and polymer synthesis to create a number of new molecular architectures (Figure 3.18).[180,181] The introduction of specific metal-based catalysts (*e.g.* the ruthenium-based Grubbs catalyst) enables the rapid interchange of olefin bonds that is an essential component of ring opening metathesis polymerization (ROMP). In a recent demonstration of bond shuffling, Otsuka *et al.*[182] utilized two homopolymers containing olefins within the polymer backbone and introduced Grubbs' catalyst into the polymer mixture. Over time, the catalyzed metathesis reaction transformed the population of polymer chains into alternating units of both homopolymers. In a subsequent study, Lu *et al.*[183] applied this concept to a crosslinked polybutadiene network containing olefins within its backbone. By adjusting the concentration of the second-generation Grubbs' catalyst, they were able to control the rate of stress relaxation. This straightforward approach of inducing network rearrangement *via* metathesis is broadly applicable to most polymer networks containing olefins.

3.5 Kinetics and Mechanical Properties

In earlier sections, the design of CANs emphasized bond connectivity and rearrangement type as primary considerations. While these factors are crucial in CAN design, understanding the microscopic dynamics is equally essential for determining the unique properties of these materials. This section will provide a brief overview of the dynamic mechanical behavior associated with CANs and establish a connection to the underlying microscopic kinetics of bond rearrangement.

The most distinctive feature of a CAN is the apparent contradiction of being a chemically crosslinked material capable of undergoing plastic deformation. It possesses a gel-point yet lacks a finite zero-frequency modulus. Because of their dynamic nature, dynamic mechanical analysis (DMA) is frequently employed to investigate their time-dependent mechanical response.

Similar to conventional covalent networks, CANs exhibit a rubbery 'plateau' modulus at intermediate frequencies, which increases sharply upon transitioning to high frequency glassy dynamics (Figure 3.19).[184] However, unlike conventional networks, these crosslinked materials exhibit characteristics of a viscoelastic liquid at low frequency (*i.e.*, $G' \sim \omega^2$ and $G'' \sim \omega^1$).[117] The dynamic mechanical behavior observed in CANs is, therefore, similar to that observed in linear polymer melts. Nevertheless, there are two critical differences between a linear polymer melt and a CAN: (1) the plateau modulus in a melt is related to the entanglements, whereas in a CAN it is related to the crosslink density, and (2) relaxation (*i.e.*, low-frequency behavior) in a melt is related to the curvilinear diffusion of a polymer chain (*i.e.*, reptation), whereas in a CAN it is related to the reverse reaction kinetics of the crosslinks.[1,117]

Adzima *et al.*[117] critically examined both the chemical kinetics and the dynamic mechanical properties of a thermoreversible Diels–Alder dissociative network. Their study yielded two key insights: (1) increasing temperature reduces the number of crosslinks and consequently lowers the plateau modulus, and (2) the longest relaxation time of the network correlates with the reverse reaction kinetics of the crosslink (*i.e.*, the Diels–Alder adduct half-life). These observations suggest that one can predict the intermediate to low frequency portion of the DMA curve given the reaction kinetics of a reversible addition monomer-type network architecture.

Associative CANs undergo cleavage and reformation in a manner that, for practical purposes, maintains the total number of crosslinks, thus minimizing the temperature sensitivity of the plateau modulus (aside from the entropic effects described by rubber elasticity theory). Nevertheless, these materials still exhibit relaxation at low frequencies (long timescales). Indeed, Montarnal *et al.*[17] observed a nearly temperature-independent modulus

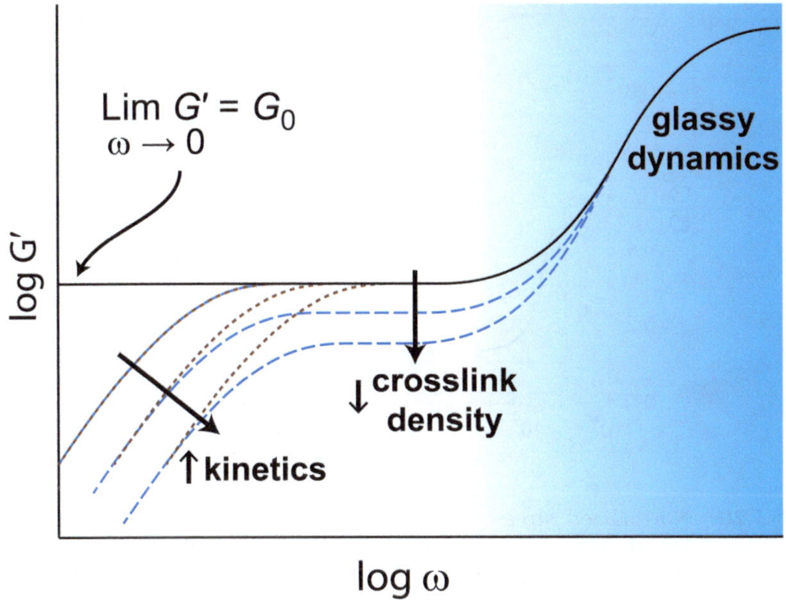

Figure 3.19 The idealized complex elastic (shear) modulus as a function of frequency is depicted for a viscoelastic solid (solid black line), dissociative CAN (dotted red line), and associative CAN (dashed blue line). Covalently crosslinked networks typically exhibit the mechanical behavior of a viscoelastic solid, where the zero-frequency modulus is directly related to the crosslink density *via* rubber elasticity theory (*i.e.*, $G_0 \sim \rho_x T$). Both associative and dissociative CANs show the relaxation behavior of a viscoelastic liquid ($G' \sim \omega^2$), directly related to the kinetics of their respective exchange reactions. While associative CANs maintain relatively constant crosslink densities under varying reaction conditions, dissociative CANs exist in equilibrium between crosslinked and uncrosslinked states, resulting in changes in their plateau modulus. All polymer networks exhibit a high-frequency modulus associated with glassy dynamics (shaded region). Adapted from ref. 7 with permission from the Royal Society of Chemistry.

(or rather a linear increase consistent with rubber elasticity theory) and a significant decrease in relaxation time with increasing temperature (Figure 3.20).

It is worth noting that the preceding discussion of mechanical behavior of associative and dissociative CANs primarily pertains to *monomer* network architectures. In contrast, relaxation behavior in *macromer* network architectures is anticipated to be complicated by the convolution of reversible crosslinks and relaxation through reptation. Despite these complexities, for such material systems some of the initial CANs showcase remolding and healing capabilities. Moreover, the tunable stability afforded by the macromer CAN design underscores their significance in both current and future materials applications.

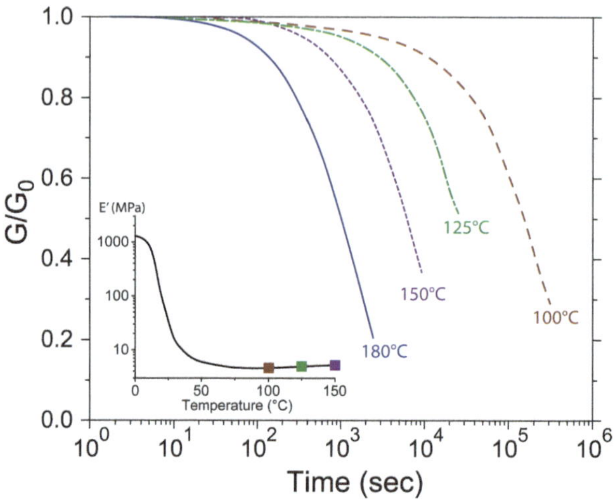

Figure 3.20 Normalized stress relaxation for an associative, transesterification-based network (see Section 3.4.3.1), the rate of which increases with increasing temperature. DMA (inset) reveals linear elastic modulus increases within increasing temperature consistent with rubber elasticity theory. Reproduced from ref. 7 with permission from the Royal Society of Chemistry.

3.6 Potential Applications of CANs

Since the publication of the first edition of this book, research into CANs has progressed from understanding the fundamental chemistries that enable healing, to demonstrating their potential to replace conventional polymeric materials in everyday objects. Although products containing CANs have yet to achieve commercialization, laboratory scale prototyping has begun to point the way towards markets that would benefit from their unique properties, as shown in the select examples set out below.

3.6.1 3D Printing

In the last couple of decades, commercial applications for 3D printing have increased significantly due to new materials processing and printing technologies. Some of the remaining challenges for 3D printed materials are poor interlayer bonding, material anisotropy, and support-structure artifacts that result from the layering printing process. All these defects can be healed using CANs. While there are several 3D printing types, two will be examined here: stereolithography (SLA) and fused filament fabrication (FFF) printing. SLA printing using photolithography to print structures layer by layer using photomasks, lasers, or digital projection, whereas FFF printing heats the material and extrudes it through a nozzle. SLA and FFF printing applications have different CAN requirements, but there are significant advantages of CANs over traditional thermoplastics and thermosets.[185]

There are several monomers that are suitable for SLA printing, typically involving a photoinitiator and a monomer capable of addition polymerization. One common example is the radical initiated polymerization of acrylates, where the dynamic covalent linkage is incorporated within the backbone of the monomer (*e.g.*, see the CAN formation strategy shown in D). Both reversible addition and degenerative exchange linkages are suitable strategies for network healing. For example, hindered urea bonds that have been used in dual cure photoresins, first described by Velankar *et al.*[186] and implemented in commercial SLA printing technologies,[187] could also be used in materials healing. In another example, Li *et al.*[188] used SLA printing of diacrylate with disulfide linkages to make 3D objects capable of being healed (Figure 3.21).

Traditional crosslinked polymers are incompatible with FFF-based 3D printing; however, thermally activated dissociative CANs are able to flow at high temperatures. Additionally, the 'ink' in FFF printing is extruded in a columnar fashion, leading to challenges associated with interfacial bonding and anisotropy. Dissociative CANs, such as those crosslinked by the furan–maleimide Diels–Alder adduct, essentially depolymerize during printing and can be covalently integrated into the subsequent layer upon cooling (Figure 3.22).[189–193]

3.6.2 CANS in Healable Footwear

In 2021, Sun and co-workers developed a crosslinked polydimethylsiloxane elastomer incorporating pyrazoles in the main chain (Figure 3.23).[194] The pyrazole ring substituents were systematically varied, with 'R' groups differing in steric size and electronic donating capacity. The resultant materials exhibited a range of properties, depending on the nature of the substituent. Notably, polymers with larger substituents on the pyrazole ring, such as $R{=}C(CH_3)_3$ (*e.g.*, AtBuPu-PDMS), facilitated dynamic exchange between pyrazole urea bonds. As a consequence of the low cost of the starting materials and the straightforward synthesis, this polymer could be produced on a kilo scale. It was formed into orthopedic insoles, which demonstrated effective crack healing, as shown in Figure 3.24.

3.6.3 CANS in Wearable Electronics

Wearable electronics have become increasingly prevalent in recent years, integrated into devices such as fitness trackers and medical sensors. You and co-workers developed a degradable and recyclable conductive polymer composite.[195] The crosslinked polymer incorporated both ester groups and Diels–Alder adducts, enabling the potential for environmental degradation and healing under appropriate conditions (Figure 3.25). The material's ability to restore conductivity after fracture was demonstrated by including a sample of the material in a functioning circuit (Figure 3.26). The pristine material conducted sufficient electricity to power LEDs, and after breaking

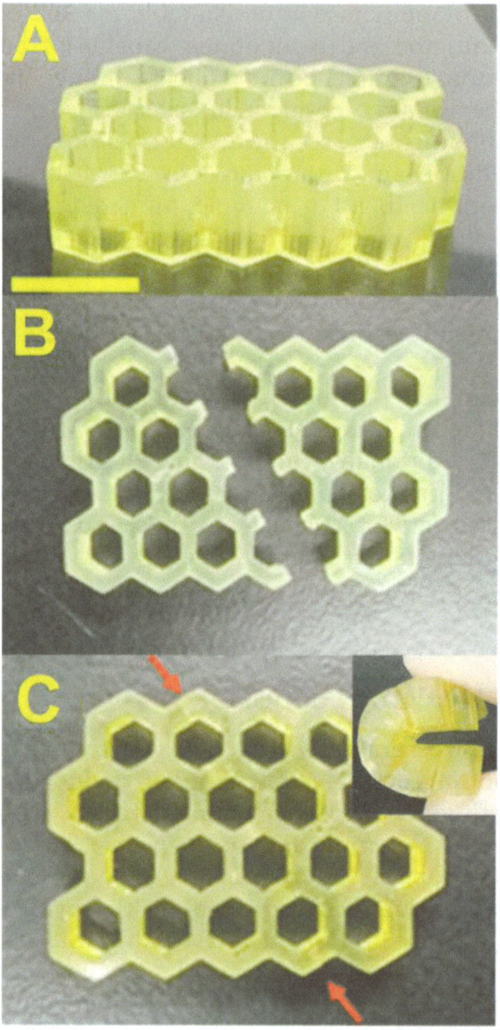

Figure 3.21 (A) SLA print of a honeycomb object using a diacrylate monomer containing disulfide bonds. The honeycomb was cut into two pieces (B) and then healed *via* heating (C) (adapted from ref. 188 with permission from the American Chemical Society, Copyright 2019).

the circuit by cutting the conducting CAN, the material could be healed through the application of IR irradiation, restoring the function of the circuit.

3.6.4 CANS as Switchable Adhesives

Adhesives are an essential part of manufacturing, enabling quick and effective bonding of diverse materials. However, this generates a challenge at

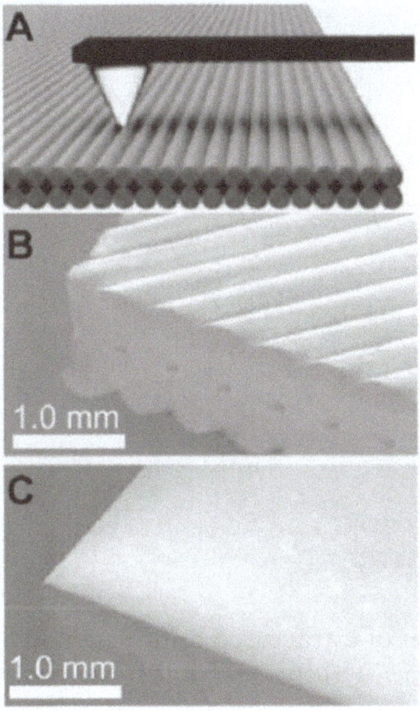

Figure 3.22 Characterization of a 3D FFF printed material using a stylus profilometer, depicted in A. 3D printing of a commercial thermoplastic polyurethane results in gaps (shown in B) that are not present in the furan–maleimide based resins (C). Adapted from ref. 189 with permission from John Wiley & Sons, Copyright 2017 Wiley-VCH Verlag GmbH & Co. KGaA, Weinheim.

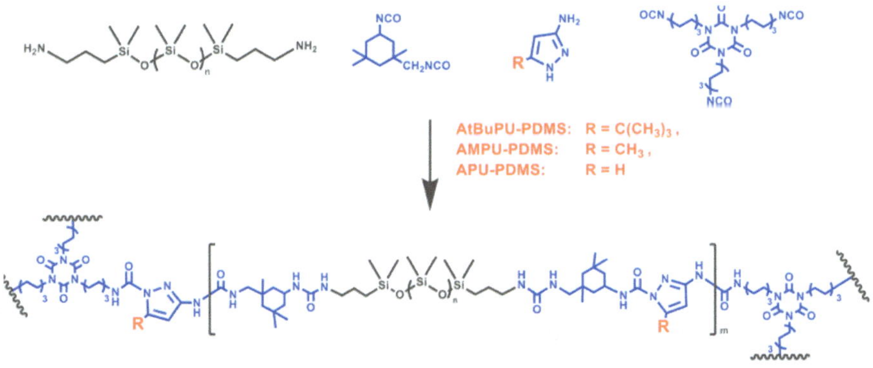

Figure 3.23 Synthesis of the CAN for footwear applications produced by Sun and co-workers (adapted from ref. 194 with permission from Elsevier, Copyright 2021).

the end of life of multi-component products, as the materials are difficult to separate and recycle. The widespread development of switchable adhesives, which can rapidly transition from a high to a low bond strength, would ameliorate many of these problems.

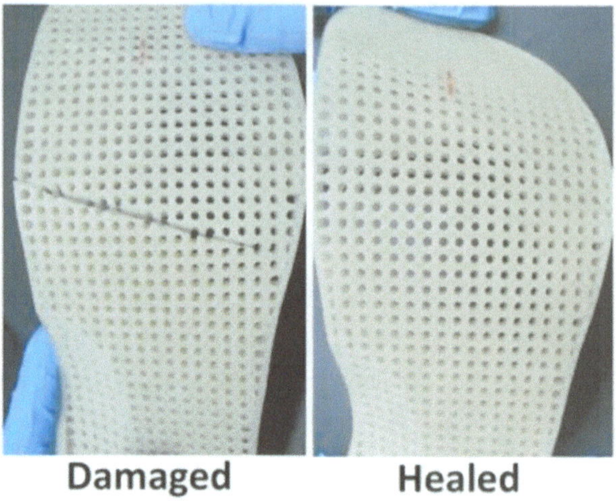

Figure 3.24 Photograph of a fractured and healed orthopedic inner sole (adapted from ref. 194 with permission from Elsevier, Copyright 2021).

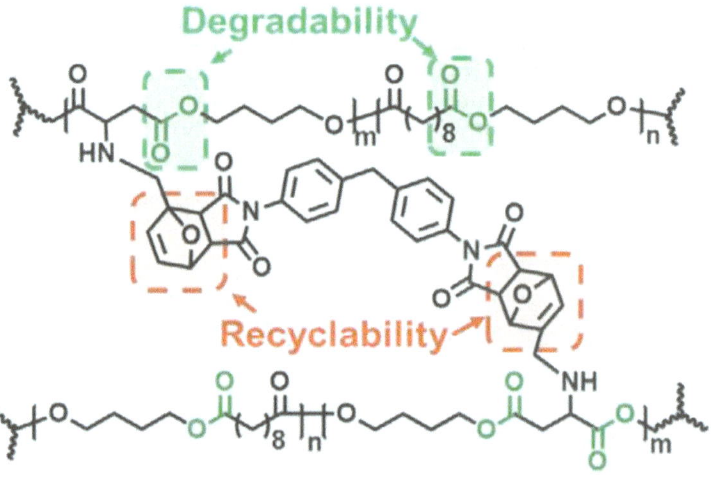

Figure 3.25 Structure of the polymeric component of the healable conductive composite produced by You and co-workers (adapted from ref. 195 with permission from John Wiley & Sons, Copyright 2020 Wiley-VCH GmbH).

Verge and co-workers produced a cardanol-based polybenzoxazine vitrimer that had reversible adhesive properties.[196] The system worked through dithiol exchange at elevated temperatures (Figure 3.27). The adhesive was able to join aluminum samples with bond strengths up to 2.8 MPa. Additionally, the system was able to undergo multiple bonding/debonding cycles with minimal loss of adhesive strength when subject to healing at 120 °C and 8 bar pressure.

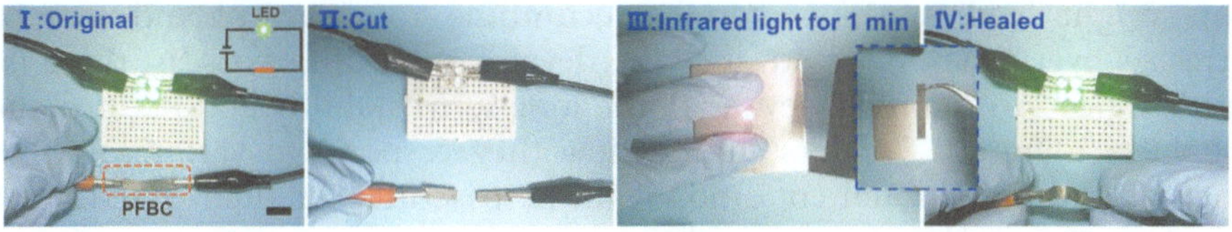

Figure 3.26 Demonstration of the restoration of conductivity by IR inducted healing of a conductive CAN. Adapted from ref. 195 with permission from John Wiley & Sons, Copyright 2020 Wiley-VCH GmbH.

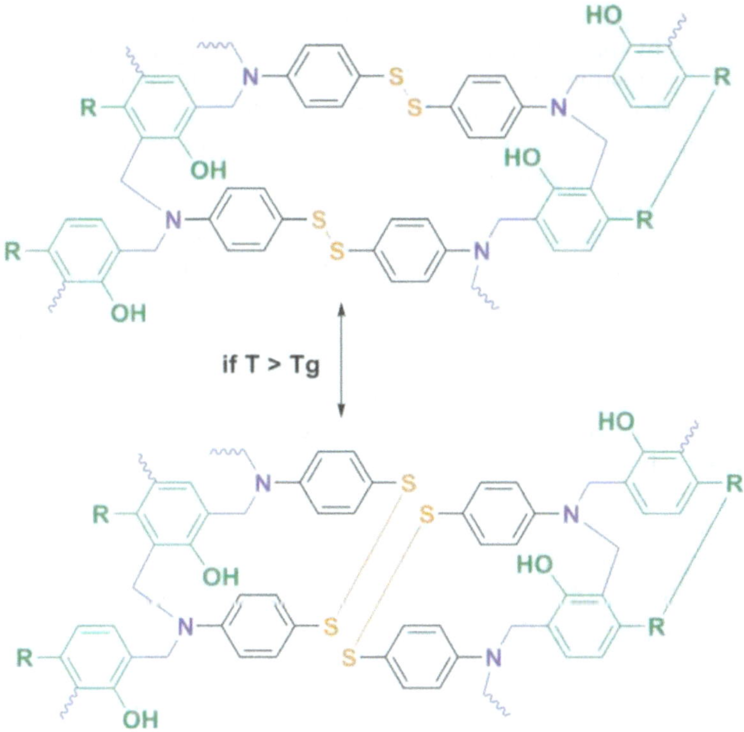

Figure 3.27 Simplified structure of the adhesive vitrimer produced by Verge and co-workers showing the proposed healing through dithiol exchange at elevated temperatures (reproduced from ref. 196 with permission from the Royal Society of Chemistry).

3.7 Conclusion and Outlook

The incorporation of healing within covalently crosslinked materials not only enables new avenues for smart materials applications but also plays a critical role in addressing the grand challenge of materials sustainability. These dynamic materials possess tremendous potential to adapt to environmental challenges. In some cases, this might lead to homogenizing deleterious stress increasing the material's service lifetime; in other cases, this might lead to the material succumbing to chemical and material stresses over time. As we advance the development of new CANs, there is a critical need to rigorously evaluate the long-term stability and performance of these materials. Understanding how they respond to various environmental and operational conditions will be essential for their practical implementation and widespread adoption.

Perhaps one of the foremost challenges in the application of CANs lies in their positioning as alternatives to commercially established thermosets. While CANs offer promising avenues for sustainable solutions, replacing these established materials poses significant challenges. Commercially available materials have a long track record, particularly concerning aging and consistent material properties over a range of applications. There is a need for new testing methodologies to systematically assess how chemical and topological designs of CANs influence relevant application properties throughout their lifecycle. Bridging this gap will be critical for validating CANs as reliable alternatives and ensuring their long-term viability in practical applications.

New implementations of reversible addition and degenerative exchange chemistries in concert with topological design will drive innovations in covalent network healing strategies. Mechanochemistry, for example, has great potential to accelerate network adaptability or initiate self-healing in polymer networks, but effective topological design is required to direct mechanical stress within the network strands. There is also a significant need to characterize reaction kinetics alongside dynamic mechanical properties to gain new insights in CAN design. Nevertheless, the foundational concepts and tools are currently available for the *a priori* design of CANs for specific healing applications. By harnessing reversible crosslinks within polymer network frameworks, a new class of smart materials has emerged, promising to extend the service lifetime of traditional crosslinked polymers in a range of materials applications.

References

1. C. J. Kloxin, T. F. Scott, B. J. Adzima and C. N. Bowman, *Macromolecules*, 2010, **43**, 2643–2653.
2. M. Capelot, D. Montarnal, F. Tournilhac and L. Leibler, *J. Am. Chem. Soc.*, 2012, **134**, 7664–7667.
3. L. P. Engle and K. B. Wagener, *J. Macromol. Sci., Rev. Macromol. Chem. Phys.*, 1993, **C33**, 239–257.
4. J. M. Lehn, *Chem. – Eur. J.*, 1999, **5**, 2455–2463.
5. M. E. Belowich and J. F. Stoddart, *Chem. Soc. Rev.*, 2012, **41**, 2003–2024.
6. S. J. Rowan, S. J. Cantrill, G. R. L. Cousins, J. K. M. Sanders and J. F. Stoddart, *Angew. Chem., Int. Ed.*, 2002, **41**, 898–952.
7. C. J. Kloxin and C. N. Bowman, *Chem. Soc. Rev.*, 2013, **42**, 7161–7173.
8. G. M. Scheutz, J. J. Lessard, M. B. Sims and B. S. Sumerlin, *J. Am. Chem. Soc.*, 2019, **141**, 16181–16196.

9. W. Denissen, J. M. Winne and F. E. Du Prez, *Chem. Sci.*, 2016, **7**, 30–38.
10. M. Guerre, C. Taplan, J. M. Winne and F. E. Du Prez, *Chem. Sci.*, 2020, **11**, 4855–4870.
11. M. K. McBride, B. T. Worrell, T. Brown, L. M. Cox, N. Sowan, C. Wang, M. Podgorski, A. M. Martinez and C. N. Bowman, *Annu. Rev. Chem. Biomol. Eng.*, 2019, **10**, 175–198.
12. S. V. Wanasinghe, O. J. Dodo and D. Konkolewicz, *Angew. Chem., Int. Ed.*, 2022, **61**, e202206938.
13. P. J. Flory, *Chem. Rev.*, 1944, **35**, 51–75.
14. M. S. Green and A. V. Tobolsky, *J. Chem. Phys.*, 1946, **14**, 80–92.
15. J. M. Craven, *USA Pat.*, 3435003, 1969.
16. J. M. Lehn, *Prog. Polym. Sci.*, 2005, **30**, 814–831.
17. D. Montarnal, M. Capelot, F. Tournilhac and L. Leibler, *Science*, 2011, **334**, 965–968.
18. C. A. Angell, *Science*, 1995, **267**, 1924–1935.
19. B. R. Elling and W. R. Dichtel, *ACS Cent. Sci.*, 2020, **6**, 1488–1496.
20. A. B. S. Bakhtiari, D. Hsiao, G. X. Jin, B. D. Gates and N. R. Branda, *Angew. Chem., Int. Ed.*, 2009, **48**, 4166–4169.
21. S. Yamashita, H. Fukushima, Y. Niidome, T. Mori, Y. Katayama and T. Niidome, *Langmuir*, 2011, **27**, 14621–14626.
22. B. J. Adzima, C. J. Kloxin and C. N. Bowman, *Adv. Mater.*, 2010, **22**, 2784–2787.
23. R. J. Wojtecki, M. A. Meador and S. J. Rowan, *Nat. Mater.*, 2011, **10**, 14–27.
24. D. Greszta, D. Mardare and K. Matyjaszewski, *Macromolecules*, 1994, **27**, 638–644.
25. K. Matyjaszewski, S. Gaynor and J.-S. Wang, *Macromolecules*, 1995, **28**, 2093–2095.
26. H. C. Kolb, M. G. Finn and K. B. Sharpless, *Angew. Chem., Int. Ed.*, 2001, **40**, 2004.
27. A. Gandini, *Prog. Polym. Sci.*, 2013, **38**, 1–29.
28. D. J. Fortman, J. P. Brutman, M. A. Hillmyer and W. R. Dichtel, *J. Appl. Polym. Sci.*, 2017, **134**, 44984.
29. D. J. Fortman, J. P. Brutman, C. J. Cramer, M. A. Hillmyer and W. R. Dichtel, *J. Am. Chem. Soc.*, 2015, **137**, 14019–14022.
30. R. L. Snyder, D. J. Fortman, G. X. De Hoe, M. A. Hillmyer and W. R. Dichtel, *Macromolecules*, 2018, **51**, 389–397.
31. B. T. Worrell, M. K. McBride, G. B. Lyon, L. M. Cox, C. Wang, S. Mavila, C.-H. Lim, H. M. Coley, C. B. Musgrave, Y. Ding and C. N. Bowman, *Nat. Commun.*, 2018, **9**, 2804.
32. M. Podgorski, S. Mavila, S. Huang, N. Spurgin, J. Sinha and C. N. Bowman, *Angew. Chem., Int. Ed.*, 2020, **59**, 9345–9349.
33. M. M. Obadia, B. P. Mudraboyina, A. Serghei, D. Montarnal and E. Drockenmuller, *J. Am. Chem. Soc.*, 2015, **137**, 6078–6083.
34. E. Ruckenstein and X. N. Chen, *Macromolecules*, 2000, **33**, 8992–9001.

35. C. F. Gibbs, E. R. Littmann and C. S. Marvel, *J. Am. Chem. Soc.*, 1933, **55**, 753–757.
36. A. Jourdain, R. Asbai, O. Anaya, M. M. Chehimi, E. Drockenmuller and D. Montarnal, *Macromolecules*, 2020, **53**, 1884–1900.
37. W. Denissen, G. Rivero, R. Nicolaÿ, L. Leibler, J. M. Winne and F. E. Du Prez, *Adv. Funct. Mater.*, 2015, **25**, 2451–2457.
38. T. Stukenbroeker, W. Wang, J. M. Winne, F. E. Du Prez, R. Nicolaÿ and L. Leibler, *Polym. Chem.*, 2017, **8**, 6590–6593.
39. J. Liu, Y. Shi, J.-J. Li, Z.-H. Luo and Y.-N. Zhou, *Macromolecules*, 2023, **56**, 6644–6654.
40. C. He, P. R. Christensen, T. J. Seguin, E. A. Dailing, B. M. Wood, R. K. Walde, K. A. Persson, T. P. Russell and B. A. Helms, *Angew. Chem., Int. Ed.*, 2020, **59**, 735–739.
41. P. R. Christensen, A. M. Scheuermann, K. E. Loeffler and B. A. Helms, *Nat. Chem.*, 2019, **11**, 442–448.
42. K. L. Diehl, I. V. Kolesnichenko, S. A. Robotham, J. L. Bachman, Y. Zhong, J. S. Brodbelt and E. V. Anslyn, *Nat. Chem.*, 2016, **8**, 968–973.
43. J. S. A. Ishibashi and J. A. Kalow, *ACS Macro Lett.*, 2018, **7**, 482–486.
44. B. M. El-Zaatari, J. S. A. Ishibashi and J. A. Kalow, *Polym. Chem.*, 2020, **11**, 5339–5345.
45. T. F. Scott, A. D. Schneider, W. D. Cook and C. N. Bowman, *Science*, 2005, **308**, 1615–1617.
46. Q. Li, S. Q. Ma, S. Wang, W. C. Yuan, X. W. Xu, B. B. Wang, K. F. Huang and J. Zhu, *J. Mater. Chem. A*, 2019, **7**, 18039–18049.
47. S. Maes, T. Habets, S. M. Fischer, B. Grignard, C. Detrembleur and F. E. Du Prez, *Polym. Chem.*, 2024, **15**, 2296–2307.
48. Q. Li, S. Q. Ma, N. Lu, J. F. Qiu, J. L. Ye, Y. L. Liu, S. Wang, Y. Y. Han, B. B. Wang, X. W. Xu, H. Z. Feng and J. Zhu, *Green Chem.*, 2020, **22**, 7769–7777.
49. T. Habets, G. Seychal, M. Caliari, J. M. Raquez, H. Sardon, B. Grignard and C. Detrembleur, *J. Am. Chem. Soc.*, 2023, **145**, 25450–25462.
50. N. Van Herck, D. Maes, K. Unal, M. Guerre, J. M. Winne and F. E. Du Prez, *Angew. Chem., Int. Ed.*, 2020, **59**, 3609–3617.
51. V. V. Zhang, J. Accardo, I. Kevlishvili, E. F. Woods, S. J. Chapman, C. T. Eckdahl, C. L. Stern, H. J. Kulik and J. A. Kalow, *Chem*, 2023, **9**, 21.
52. A. R. Gilbert and S. W. Kantor, *J. Polym. Sci.*, 1959, **40**, 35–58.
53. S. W. Kantor, W. T. Grubb and R. C. Osthoff, *J. Am. Chem. Soc.*, 1954, **76**, 5190–5197.
54. P. W. Zheng and T. J. McCarthy, *J. Am. Chem. Soc.*, 2012, **134**, 2024–2027.
55. R. C. Osthoff, A. M. Bueche and W. T. Grubb, *J. Am. Chem. Soc.*, 1954, **76**, 4659–4663.
56. R. W. Layer, *Chem. Rev.*, 1963, **63**, 489–510.
57. C. D. Meyer, C. S. Joiner and J. F. Stoddart, *Chem. Soc. Rev.*, 2007, **36**, 1705–1723.
58. S. Dhers, G. Vantomme and L. Avérous, *Green Chem.*, 2019, **21**, 1596–1601.

59. J. Y. Chang, S. K. Do and M. J. Han, *Polymer*, 2001, **42**, 7589–7594.
60. B. Zhang, Z. A. Digby, J. A. Flum, P. Chakma, J. M. Saul, J. L. Sparks and D. Konkolewicz, *Macromolecules*, 2016, **49**, 6871–6878.
61. D. Berne, V. Ladmiral, E. Leclerc and S. Caillol, *Polymers*, 2022, **14**, 4457.
62. Y. Higaki, H. Otsuka and A. Takahara, *Macromolecules*, 2006, **39**, 2121–2125.
63. Y. Amamoto, M. Kikuchi, H. Masunaga, S. Sasaki, H. Otsuka and A. Takahara, *Macromolecules*, 2009, **42**, 8733–8738.
64. M. Nagata and Y. Yamamoto, *J. Polym. Sci. Pol. Chem.*, 2009, **47**, 2422–2433.
65. Y. Jia, G. Delaittre and M. Tsotsalas, *Macromol. Mater. Eng.*, 2022, **307**, 2200178.
66. J. J. Cash, T. Kubo, A. P. Bapat and B. S. Sumerlin, *Macromolecules*, 2015, **48**, 2098–2106.
67. M. Röttger, T. Domenech, R. van der Weegen, A. Breuillac, R. Nicolaÿ and L. Leibler, *Science*, 2017, **356**, 62–65.
68. W. A. Ogden and Z. Guan, *J. Am. Chem. Soc.*, 2018, **140**, 6217–6220.
69. S. R. Trenor, A. R. Shultz, B. J. Love and T. E. Long, *Chem. Rev.*, 2004, **104**, 3059–3077.
70. M. Nagata and Y. Yamamoto, *React. Funct. Polym.*, 2008, **68**, 915–921.
71. D. L. Zhao, B. Y. Ren, S. S. Liu, X. X. Liu and Z. Tong, *Chem. Commun.*, 2006, 779–781.
72. B. Y. Ren, D. L. Zhao, S. S. Liu, X. X. Liu and Z. Tong, *Macromolecules*, 2007, **40**, 4501–4508.
73. A. Lendlein, H. Y. Jiang, O. Junger and R. Langer, *Nature*, 2005, **434**, 879–882.
74. H. Y. Jiang, S. Kelch and A. Lendlein, *Adv. Mater.*, 2006, **18**, 1471–1475.
75. Y. J. Zheng, F. M. Andreopoulos, M. Micic, Q. Huo, S. M. Pham and R. M. Leblanc, *Adv. Funct. Mater.*, 2001, **11**, 37–40.
76. M. Micic, Y. J. Zheng, V. Moy, X. H. Zhang, F. M. Andreopoulos and R. M. Leblanc, *Colloids Surf., B*, 2003, **27**, 147–158.
77. K. M. Gattas-Asfura, E. Weisman, F. M. Andreopoulos, M. Micic, B. Muller, S. Sirpal, S. M. Pham and R. M. Leblanc, *Biomacromolecules*, 2005, **6**, 1503–1509.
78. Y. Imai, T. Ogoshi, K. Naka and Y. Chujo, *Polym. Bull.*, 2000, **45**, 9–16.
79. Y. J. Zheng, M. Mieie, S. V. Mello, M. Mabrouki, F. M. Andreopoulos, V. Konka, S. M. Pham and R. M. Leblanc, *Macromolecules*, 2002, **35**, 5228–5234.
80. L. A. Connal, R. Vestberg, C. J. Hawker and G. G. Qiao, *Adv. Funct. Mater.*, 2008, **18**, 3315–3322.
81. J. Matsui, Y. Ochi and K. Tamaki, *Chem. Lett.*, 2006, **35**, 80–81.
82. A. V. Tobolsky, W. J. Macknight and M. Takahashi, *J. Phys. Chem.*, 1964, **68**, 787–790.
83. Y. Takahash and A. V. Tobolsky, *Polym. J.*, 1971, **2**, 457–467.
84. J. A. Yoon, J. Kamada, K. Koynov, J. Mohin, R. Nicolay, Y. Z. Zhang, A. C. Balazs, T. Kowalewski and K. Matyjaszewski, *Macromolecules*, 2012, **45**, 142–149.

85. G. H. Deng, F. Y. Li, H. X. Yu, F. Y. Liu, C. Y. Liu, W. X. Sun, H. F. Jiang and Y. M. Chen, *ACS Macro Lett.*, 2012, **1**, 275–279.
86. B. D. Fairbanks, S. P. Singh, C. N. Bowman and K. S. Anseth, *Macromolecules*, 2011, **44**, 2444–2450.
87. J. Canadell, H. Goossens and B. Klumperman, *Macromolecules*, 2011, **44**, 2536–2541.
88. S. Cerritelli, D. Velluto and J. A. Hubbell, *Biomacromolecules*, 2007, **8**, 1966–1972.
89. M. Pooga, U. Soomets, M. Hallbrink, A. Valkna, K. Saar, K. Rezaei, U. Kahl, J. X. Hao, X. J. Xu, Z. Wiesenfeld-Hallin, T. Hokfelt, T. Bartfai and U. Langel, *Nat. Biotechnol.*, 1998, **16**, 857–861.
90. G. Saito, J. A. Swanson and K. D. Lee, *Adv. Drug Delivery Rev.*, 2003, **55**, 199–215.
91. S. Takae, K. Miyata, M. Oba, T. Ishii, N. Nishiyama, K. Itaka, Y. Yamasaki, H. Koyama and K. Kataoka, *J. Am. Chem. Soc.*, 2008, **130**, 6001–6009.
92. O. Ramstrom and J. M. Lehn, *ChemBioChem*, 2000, **1**, 41–48.
93. G. Dalman, J. McDermed and G. Gorin, *J. Org. Chem.*, 1964, **29**, 1480.
94. A. Fava, A. Iliceto and E. Camera, *J. Am. Chem. Soc.*, 1957, **79**, 833–838.
95. J. Houk and G. M. Whitesides, *J. Am. Chem. Soc.*, 1987, **109**, 6825–6836.
96. N. V. Tsarevsky and K. Matyjaszewski, *Macromolecules*, 2002, **35**, 9009–9014.
97. Y. Chujo, K. Sada, A. Naka, R. Nomura and T. Saegusa, *Macromolecules*, 1993, **26**, 883–887.
98. R. Caraballo, M. Rahm, P. Vongvilai, T. Brinck and O. Ramstrom, *Chem. Commun.*, 2008, 6603–6605.
99. R. Caraballo, M. Sakulsombat and O. Ramstrom, *Chem. Commun.*, 2010, **46**, 8469–8471.
100. A. Rekondo, R. Martin, A. Ruiz de Luzuriaga, G. Cabañero, H. J. Grande and I. Odriozola, *Mater. Horiz.*, 2014, **1**, 237–240.
101. R. Martin, A. Rekondo, A. Ruiz de Luzuriaga, G. Cabañero, H. J. Grande and I. Odriozola, *J. Mater. Chem. A*, 2014, **2**, 5710–5715.
102. R. C. Boutelle and B. H. Northrop, *J. Org. Chem.*, 2011, **76**, 7994–8002.
103. K. C. Koehler, A. Durackova, C. J. Kloxin and C. N. Bowman, *AIChE J.*, 2012, **58**, 3545–3552.
104. F. Fringuelli and A. Taticchi, *The Diels–Alder Reaction: Selected Practical Methods*, John Wiley & Sons, Inc., New York, 2002.
105. H. Laita, S. Boufi and A. Gandini, *Eur. Polym. J.*, 1997, **33**, 1203–1211.
106. C. Gousse, A. Gandini and P. Hodge, *Macromolecules*, 1998, **31**, 314–321.
107. R. Gheneim, C. Perez-Berumen and A. Gandini, *Macromolecules*, 2002, **35**, 7246–7253.
108. E. Goiti, F. Heatley, M. B. Huglin and J. M. Rego, *Eur. Polym. J.*, 2004, **40**, 1451–1460.
109. E. Goiti, M. B. Huglin and J. M. Rego, *Eur. Polym. J.*, 2004, **40**, 219–226.
110. Y. L. Liu and Y. W. Chen, *Macromol. Chem. Phys.*, 2007, **208**, 224–232.

111. Y. L. Liu and C. Y. Hsieh, *J. Polym. Sci. Pol. Chem.*, 2006, **44**, 905–913.
112. A. A. Kavitha and N. K. Singha, *Macromol. Chem. Phys.*, 2007, **208**, 2569–2577.
113. Y. Zhang, A. A. Broekhuis and F. Picchioni, *Macromolecules*, 2009, **42**, 1906–1912.
114. Y. Imai, H. Itoh, K. Naka and Y. Chujo, *Macromolecules*, 2000, **33**, 4343–4346.
115. X. X. Chen, M. A. Dam, K. Ono, A. Mal, H. B. Shen, S. R. Nutt, K. Sheran and F. Wudl, *Science*, 2002, **295**, 1698–1702.
116. X. X. Chen, F. Wudl, A. K. Mal, H. B. Shen and S. R. Nutt, *Macromolecules*, 2003, **36**, 1802–1807.
117. B. J. Adzima, H. A. Aguirre, C. J. Kloxin, T. F. Scott and C. N. Bowman, *Macromolecules*, 2008, **41**, 9112–9117.
118. J. P. Kennedy and K. F. Castner, *J. Polym. Sci. Pol. Chem.*, 1979, **17**, 2055–2070.
119. E. B. Murphy, E. Bolanos, C. Schaffner-Hamann, F. Wudl, S. R. Nutt and M. L. Auad, *Macromolecules*, 2008, **41**, 5203–5209.
120. P. J. Boul, P. Reutenauer and J. M. Lehn, *Org. Lett.*, 2005, **7**, 15–18.
121. P. Reutenauer, P. J. Boul and J. M. Lehn, *Eur. J. Org. Chem.*, 2009 1691–1697.
122. A. J. Inglis, S. Sinnwell, M. H. Stenzel and C. Barner-Kowollik, *Angew. Chem., Int. Ed.*, 2009, **48**, 2411–2414.
123. A. Rembaum, W. Baumgartner and A. Eisenberg, *J. Polym. Sci., Part B*, 1968, **6**, 159–171.
124. S. R. Williams and T. E. Long, *Prog. Polym. Sci.*, 2009, **34**, 762–782.
125. C. M. Leir and J. E. Stark, *J. Appl. Polym. Sci.*, 1989, **38**, 1535–1547.
126. P. Chakma, Z. A. Digby, M. P. Shulman, L. R. Kuhn, C. N. Morley, J. L. Sparks and D. Konkolewicz, *ACS Macro Lett.*, 2019, **8**, 95–100.
127. P. Chakma, C. N. Morley, J. L. Sparks and D. Konkolewicz, *Macromolecules*, 2020, **53**, 1233–1244.
128. M. Hayashi and L. Chen, *Polym. Chem.*, 2020, **11**, 1713–1719.
129. L. L. Robinson, E. S. Taddese, J. L. Self, C. M. Bates, J. Read de Alaniz, Z. Geng and C. J. Hawker, *Macromolecules*, 2022, **55**, 9780–9789.
130. Z. Tang, Y. Liu, Q. Huang, J. Zhao, B. Guo and L. Zhang, *Green Chem.*, 2018, **20**, 5454–5458.
131. B. Hendriks, J. Waelkens, J. M. Winne and F. E. Du Prez, *ACS Macro Lett.*, 2017, **6**, 930–934.
132. V. Scholiers, B. Hendriks, S. Maes, T. Debsharma, J. M. Winne and F. E. Du Prez, *Macromolecules*, 2023, **56**, 9559–9569.
133. M. M. Obadia, A. Jourdain, P. Cassagnau, D. Montarnal and E. Drockenmuller, *Adv. Funct. Mater.*, 2017, **27**, 1703258.
134. A. Ausin, I. Eguiazabal, M. E. Munoz, J. J. Pena and A. Santamaria, *Polym. Eng. Sci.*, 1987, **27**, 529–534.
135. P. J. Flory, *J. Am. Chem. Soc.*, 1940, **62**, 2255–2261.
136. P. J. Flory, *J. Am. Chem. Soc.*, 1940, **62**, 2261–2264.
137. R. S. Porter and L. H. Wang, *Polymer*, 1992, **33**, 2019–2030.

138. A. Demongeot, S. J. Mougnier, S. Okada, C. Soulié-Ziakovic and F. Tournilhac, *Polym. Chem.*, 2016, **7**, 4486–4493.
139. J. P. Brutman, P. A. Delgado and M. A. Hillmyer, *ACS Macro Lett.*, 2014, **3**, 607–610.
140. J. L. Self, N. D. Dolinski, M. S. Zayas, J. Read de Alaniz and C. M. Bates, *ACS Macro Lett.*, 2018, **7**, 817–821.
141. F. I. Altuna, V. Pettarin and R. J. J. Williams, *Green Chem.*, 2013, **15**, 3360–3366.
142. I. M. Serafimova, M. A. Pufall, S. Krishnan, K. Duda, M. S. Cohen, R. L. Maglathlin, J. M. McFarland, R. M. Miller, M. Frödin and J. Taunton, *Nat. Chem. Biol.*, 2012, **8**, 471–476.
143. Y. Spiesschaert, J. Danneels, N. Van Herck, M. Guerre, G. Acke, J. Winne and F. Du Prez, *Macromolecules*, 2021, **54**, 7931–7942.
144. X. Guo, F. Gao, F. Chen, J. Zhong, L. Shen, C. Lin and Y. Lin, *ACS Macro Lett.*, 2021, **10**, 1186–1190.
145. F. Chen, F. Gao, X. Guo, L. Shen and Y. Lin, *Macromolecules*, 2022, **55**, 10124–10133.
146. Y. Ma, X. Jiang, J. Yin, C. Weder, J. A. Berrocal and Z. Shi, *Angew. Chem., Int. Ed.*, 2023, **62**, e202212870.
147. D. Colombani and P. Chaumont, *Prog. Polym. Sci.*, 1996, **21**, 439–503.
148. D. N. Hall, *J. Org. Chem.*, 1967, **32**, 2082–2087.
149. D. N. Hall, A. A. Oswald and K. Griesbau, *J. Org. Chem.*, 1965, **30**, 3829–3834.
150. T. Migita, M. Kosugi, K. Takayama and Y. Nakagawa, *Tetrahedron*, 1973, **29**, 51–55.
151. D. H. R. Barton and D. Crich, *J. Chem. Soc., Perkin Trans. 1*, 1986, 1613–1619.
152. G. F. Meijs, E. Rizzardo and S. H. Thang, *Macromolecules*, 1988, **21**, 3122–3124.
153. G. F. Meijs and E. Rizzardo, *Makromol. Chem., Rapid Commun.*, 1988, **9**, 547–551.
154. P. Cacioli, D. G. Hawthorne, R. L. Laslett, E. Rizzardo and D. H. Solomon, *J. Macromol. Sci., Chem.*, 1986, **A23**, 839–852.
155. J. Chiefari, Y. K. Chong, F. Ercole, J. Krstina, J. Jeffery, T. P. T. Le, R. T. A. Mayadunne, G. F. Meijs, C. L. Moad, G. Moad, E. Rizzardo and S. H. Thang, *Macromolecules*, 1998, **31**, 5559–5562.
156. R. A. Evans and E. Rizzardo, *J. Polym. Sci., Polym. Chem.*, 2001, **39**, 202–215.
157. R. A. Evans and E. Rizzardo, *Macromolecules*, 1996, **29**, 6983–6989.
158. R. A. Evans, G. Moad, E. Rizzardo and S. H. Thang, *Macromolecules*, 1994, **27**, 7935–7937.
159. C. J. Kloxin, T. F. Scott and C. N. Bowman, *Macromolecules*, 2009, **42**, 2551–2556.
160. H. Y. Park, C. J. Kloxin, M. F. Fordney and C. N. Bowman, *Macromolecules*, 2012, **45**, 5647–5652.
161. H. Y. Park, C. J. Kloxin, A. S. Abuelyaman, J. D. Oxman and C. N. Bowman, *Macromolecules*, 2012, **45**, 5640–5646.

162. H. Y. Park, C. J. Kloxin, T. F. Scott and C. N. Bowman, *Macromolecules*, 2010, **43**, 10188–10190.
163. H. Y. Park, C. J. Kloxin, T. F. Scott and C. N. Bowman, *Dent. Mater.*, 2010, **26**, 1010–1016.
164. C. J. Kloxin, T. F. Scott, H. Y. Park and C. N. Bowman, *Adv. Mater.*, 2011, **23**, 1977–1981.
165. S. J. Ma, S. J. Mannino, N. J. Wagner and C. J. Kloxin, *ACS Macro Lett.*, 2013, **2**, 474–477.
166. Y. Amamoto, J. Kamada, H. Otsuka, A. Takahara and K. Matyjaszewski, *Angew. Chem., Int. Ed.*, 2011, **50**, 1660–1663.
167. R. Nicolay, J. Kamada, A. Van Wassen and K. Matyjaszewski, *Macromolecules*, 2010, **43**, 4355–4361.
168. R. T. A. Mayadunne, J. Jeffery, G. Moad and E. Rizzardo, *Macromolecules*, 2003, **36**, 1505–1513.
169. G. Moad, J. Chiefari, Y. K. Chong, J. Krstina, R. T. A. Mayadunne, A. Postma, E. Rizzardo and S. H. Thang, *Polym. Int.*, 2000, **49**, 993–1001.
170. C. Boyer, V. Bulmus and T. P. Davis, *Macromol. Rapid Commun.*, 2009, **30**, 493–497.
171. S. R. Gondi, A. P. Vogt and B. S. Sumerlin, *Macromolecules*, 2007, **40**, 474–481.
172. M. Hales, C. Barner-Kowollik, T. P. Davis and M. H. Stenzel, *Langmuir*, 2004, **20**, 10809–10817.
173. Y. Amamoto, H. Otsuka, A. Takahara and K. Matyjaszewski, *ACS Macro Lett.*, 2012, **1**, 478–481.
174. Y. Z. You, C. Y. Hong, R. K. Bai, C. Y. Pan and J. Wang, *Macromol. Chem. Phys.*, 2002, **203**, 477–483.
175. Y. Amamoto, H. Otsuka, A. Takahara and K. Matyjaszewski, *Adv. Mater.*, 2012, **24**, 3975–3980.
176. G. Moad, E. Rizzardo and S. H. Thang, *Aust. J. Chem.*, 2009, **62**, 1402–1472.
177. G. Moad, E. Rizzardo and S. H. Thang, *Aust. J. Chem.*, 2006, **59**, 669–692.
178. G. Moad, E. Rizzardo and S. H. Thang, *Aust. J. Chem.*, 2005, **58**, 379–410.
179. G. Moad, E. Rizzardo and S. H. Thang, *Polymer*, 2008, **49**, 1079–1131.
180. R. H. Grubbs, *Tetrahedron*, 2004, **60**, 7117–7140.
181. R. H. Grubbs and S. Chang, *Tetrahedron*, 1998, **54**, 4413–4450.
182. H. Otsuka, T. Muta, M. Sakada, T. Maeda and A. Takahara, *Chem. Commun.*, 2009, 1073–1075.
183. Y. X. Lu, F. Tournilhac, L. Leibler and Z. B. Guan, *J. Am. Chem. Soc.*, 2012, **134**, 8424–8427.
184. M. Rubinstein and R. H. Colby, *Polymer Physics*, Oxford University Press, Inc., New York, 2003.
185. V. Bijalwan, S. Rana, G. J. Yun, K. P. Singh, M. Jamil and S. Schlögl, *Polym. Rev.*, 2024, **64**, 36–79.
186. S. Velankar, J. Pazos and S. L. Cooper, *J. Appl. Polym. Sci.*, 1996, **62**, 1361–1376.
187. J. P. Rolland, K. Chen, J. Poelma, J. Goodrich, R. Pinschmidt, J. M. DeSimone and L. M. Robeson, *USA Pat.*, 9453142, 2016.

188. X. Li, R. Yu, Y. He, Y. Zhang, X. Yang, X. Zhao and W. Huang, *ACS Macro Lett.*, 2019, **8**, 1511–1516.
189. K. Yang, J. C. Grant, P. Lamey, A. Joshi-Imre, B. R. Lund, R. A. Smaldone and W. Voit, *Adv. Funct. Mater.*, 2017, **27**, 1700318.
190. G. A. Appuhamillage, J. C. Reagan, S. Khorsandi, J. R. Davidson, W. Voit and R. A. Smaldone, *Polym. Chem.*, 2017, **8**, 2087–2092.
191. J. R. Davidson, G. A. Appuhamillage, C. M. Thompson, W. Voit and R. A. Smaldone, *ACS Appl. Mater. Interfaces*, 2016, **8**, 16961–16966.
192. T. Yuan, L. Zhang, T. Li, R. Tu and H. A. Sodano, *Polym. Chem.*, 2020, **11**, 6441–6452.
193. D. J. Bischoff and M. E. Mackay, *ACS Appl. Polym. Mater.*, 2024, **6**, 3335–3341.
194. S. Sun, G. Fei, X. Wang, M. Xie, Q. Guo, D. Fu, Z. Wang, H. Wang, G. Luo and H. Xia, *Chem. Eng. J.*, 2021, **412**, 128675.
195. Y. Guo, S. Chen, L. Sun, L. Yang, L. Zhang, J. Lou and Z. You, *Adv. Funct. Mater.*, 2021, **31**, 2009799.
196. A. Trejo-Machin, L. Puchot and P. Verge, *Polym. Chem.*, 2020, **11**, 7026–7034.

CHAPTER 4

Supramolecular Healable Materials

BARNABY W. GREENLAND*[a] AND WAYNE HAYES*[b]

[a] Department of Chemistry, University of Sussex, Falmer Campus, Sussex, BN1 9RH, UK; [b] Department of Chemistry, The University of Reading, Whiteknights, Reading, RG6 6AD, UK
*Emails: b.w.greenland@sussex.ac.uk; w.c.hayes@reading.ac.uk

4.1 Introduction to Polymers and Healing Materials

Since the introduction of Bakelite, the first commercialized fully synthetic polymer in the early 1900s, products made entirely from, or containing, polymeric materials have infiltrated every area of modern life.[1] The ubiquity of polymers stems not just from their relatively low cost, but also because of their intrinsic properties such as low density and corrosion resistance. In addition, over the last 100 years, synthetic chemists have synthesized a diverse range of polymers that vary by monomer feedstock, molecular weight, dispersity and level of branching amongst other variables to produce new materials with an ever-expanding range of properties that have been used to build our modern world.

Whatever the application of the polymeric material, be it a disposable plastic bag, or safety critical component of aircraft wing, all polymers are subject to a range of disparate environmental conditions. These may include UV radiation, extremes of temperature or repeated impacts which can cause degradation of the polymer chains and/or crack propagation. An accumulation of these external events may ultimately cause the polymeric component to fail.[2] To overcome this problem and extend the life of the product without replacing the component, researchers have developed a broad range

Polymer Chemistry Series No. 41
Healable Polymer Systems: Fundamentals, Synthesis and Applications, 2nd Edition
Edited by Barnaby W. Greenland and Wayne Hayes

Published by the Royal Society of Chemistry, www.rsc.org

of approaches to produce polymers which are able to regain their strength after damage. These materials can be broadly categorized as those that repair themselves autonomously, termed *self-healing materials*, or those that repair when subject to an external stimulus, for example heat, light, pressure or mechanical stress, referred to as *healable materials*.[3]

Polymeric materials that have properties equal to the current state-of-the-art, yet are able to regain their strength *in situ* will clearly offer extended lifetimes, additional safety advantages as well as reduced cost and potential for waste when compared to non-healing predecessors. Such properties are clearly exciting to industry from a purely economic viewpoint. Furthermore, with increased attention placed on environmental concerns over recent years, being able to repair components without replacement is clearly a significant advantage in terms of developing a more environmentally stainable economy.

A conceptually simple method to heal thermoplastic (non-crosslinked) polymers is to expose them to temperatures above the glass transition temperature (T_g) of the polymer.[4] At these elevated temperatures interdiffusion of polymer chains at fracture interfaces causes any cracks to disappear. However, the rate of interdiffusion slows with increasing molecular weight, which often results in unpractically long healing times under viable heating conditions.[5]

In a similar manner, addition of a solvent to the fracture zone to either dissolve or plasticize the polymer chains at the interface and, therefore, facilitate interdiffusion, can also lead to damage repair once the solvent has evaporated.[6] Although feasible, this methodology is frequently considered non-optimal as a consequence of a number of factors: for example, inaccessibility of the fracture site to solvent, change of dimensions of the sample after dissolution/drying or the need for toxic solvents, amongst other factors.

Moving away from using heat and externally added solvents to induce healing, early examples of healing materials used a fracture, or stress induced healing stimulus[7] (for more examples, see Chapter 2 of this book). During this process the propagation of the crack initiates the healing mechanism and as such represents the first designed methodology to produce healable polymeric materials. To achieve this, capsules incorporating a monomer were dispersed throughout the bulk polymer. When subject to force sufficient to cause a fracture, the capsules in the fracture zone rupture, releasing monomer into the void. On contact with an appropriate catalyst which has been co-dispersed in the polymer, the monomer polymerizes, *healing* the fracture.[8] In contrast to the thermal and solvent based methods detailed above, this process requires no external intervention and, therefore, can be considered an example of *self-healing.* A drawback of this system is that the monomer is consumed on polymerization and, therefore, repeated healing of the same fracture zone is not possible, although more recent studies into creating interconnect vascular systems for delivering the monomer/and or initiators can alleviate this issue (see Chapter 2).[9]

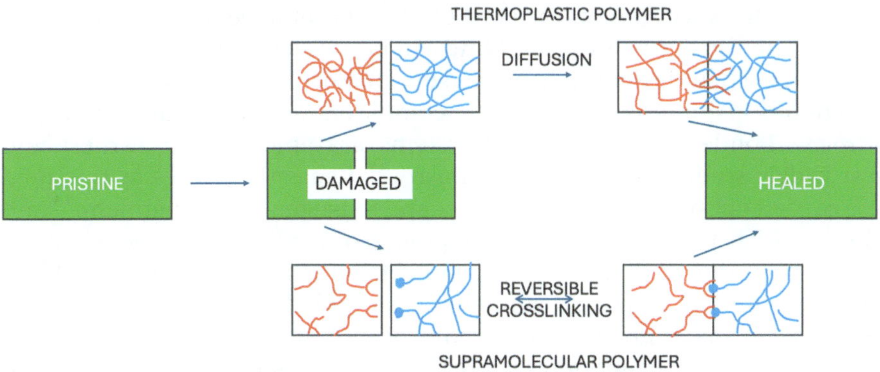

Figure 4.1 Schematic of the healing processes required for thermoplastic polymers (top) and supramolecular polymers (bottom). The state of the polymer, from pristine to damaged to healed is shown as the green blocks in the center of the figure.

Supramolecular polymeric[10] systems offer a conceptually distinctive approach to healing polymeric materials. In contrast to healing of thermoplastics where strength is regained through diffusion and re-entanglement (no new covalent bonds formed), or the encapsulation approach where new covalent bonds are formed during the polymerization reaction, this class of polymer contains supramolecular bonds which can reversibly break and reform. The strength of the supramolecular bonds can be modulated by application of an external stimuli.[3a]

Supramolecular polymers are typically constructed of monomeric or oligomeric species which self-assemble into pseudo higher molecular weight entities through the interactions between residues within which associate through supramolecular interactions.[11] These supramolecular interactions are typically weaker than the covalent bonds in the molecular species and, therefore, are the first to break when a force is applied. In contrast to covalent bonds, the supramolecular bonds can re-form and in principle, the strength of the material can be regained. Moreover, the low molecular weight species that make up the supramolecular system are able to diffuse much more rapidly than high molecular weight polymers, and therefore application of a suitable low energy stimulus (*e.g.* heat, light) can cause rapid regain of strength. The contrast between the fracture healing in linear thermoplastics and supramolecular polymers is shown schematically in Figure 4.1.

The following section sets the scene for the realization of supramolecular healable materials by providing a brief introduction to the design and realization of supramolecular systems.

4.1.1 Supramolecular Interactions: A Concise History

Supramolecular bonds are non-covalent interactions between molecules which allow them to self-assemble into larger assemblies. Through careful design of the specific components of the supramolecular assembly, the strength and directionality of the bond can be readily altered, which is of key importance when designing healable supramolecular systems.

No doubt inspired by the exquisite complex assemblies found in nature, hydrogen bonded supramolecular systems are amongst the earliest and most widely studied systems in chemistry.[12] Key early studies in this field were carried out by the groups of Meijer and Sijbesma,[13] Zimmermann[14] and Lehn,[15] which resulted in the development of synthetically accessible supramolecular motifs that can be appended to a range of oligomeric building blocks (Figure 4.2). However, supramolecular polymeric systems have since been realized that employ a broad range of such interactions, of which π–π stacking and metal–ligand interactions are also well-studied in the context of supramolecular healable polymeric materials.

Aromatic π–π stacking[16] occurs between both functionalized and unfunctionalized aromatic residues and can drive the crystal packing of simple benzene molecules in the solid state. However, the strength of the interaction can be enhanced greatly if the system is designed to include pairs of aromatic residues that have complimentary electronic properties. Typically, this means the interaction occurs between a *donor* component that is π-electron rich (conventionally colored red in structures within schematics), and an *acceptor* component that is π-electron deficient in nature (usually colored blue in structures within schematics).

Metal–ligand interactions form the bedrock of organometallic chemistry. However, the foundations for supramolecular chemistry were also laid with the breakthrough in understanding of selective metal binding by Lehn,[17] Pederson[18] and Cram.[19] The predictability of the geometry of the ligands around the metal center has enabled the synthesis of complex multicomponent structures, as exemplified by the 'grid' system produced by the Lehn group shown below containing 16 metal ions and 8 large polydentate organic ligands (Figure 4.3).[20]

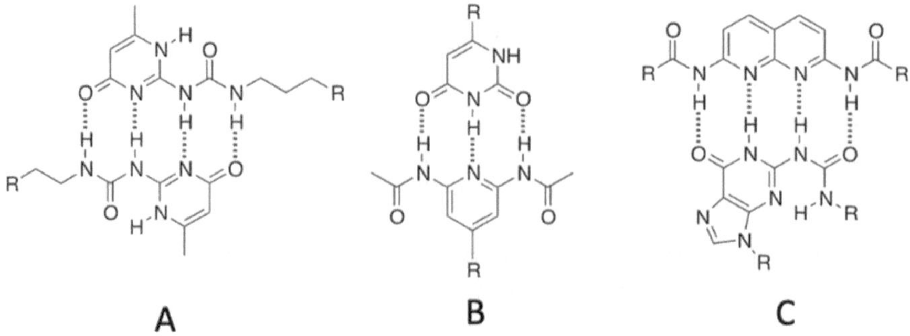

Figure 4.2 Structures of the supramolecular motifs developed by the groups of Meijer and Sijbesma, Zimmermann and Lehn (A to C, respectively). Hydrogen bonding interactions are shown in blue dashes.

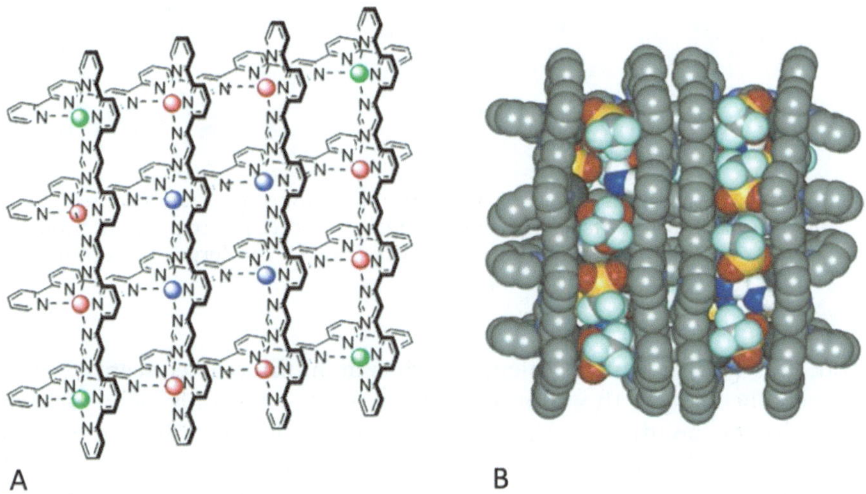

Figure 4.3 Chemical (A) and solid-state space filling structure (B) of the 4 × 4 'grid' self-assembled by Lehn and co-workers. Adapted from ref. 20 with permission from the American Chemical Society, Copyright 2003.

Central to the success of synthesis is the reversibility of the metal–ligand interactions which enables error correction during the assembly of the supramolecular complex. The system can explore series of local energy minimum without becoming kinetically trapped enabling complexes to cycle through to the lowest energy, targeted structure. This process, which mimics self-assembly processes found in nature, neatly demonstrates the key reversibility features of supramolecular systems which allow them to be used in healable materials.

4.1.2 Supramolecular Polymers: Concepts and Properties

The examples of supramolecular systems shown in the previous section were designed to produce discrete complexes, be they dimers or higher order structures. Conceptually, it is clear that the addition of complimentary, self-complimentary supramolecular motifs or mixtures of these to either end of a short oligomeric structure will result in supramolecular polymerizations, as shown in Figure 4.4.[21]

Initial studies into supramolecular polymers focused on understanding their properties in dilute solution (≫10% w/v). In this low concentration regime, it was established that the degree of polymerization (DP) for the supramolecular polymer was related to the molar concentration ([M]) of the solution and the binding constant (K_a) of the supramolecular motif according to eqn (4.1)

$$\mathrm{DP} \approx (K_a[\mathrm{M}])^{0.5} \tag{4.1}$$

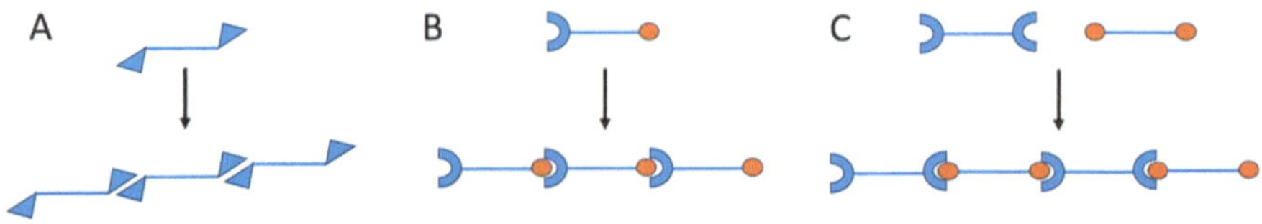

Figure 4.4 Schematics showing how ditopic 'monomers' containing (A) self-complimentary; (B) complimentary or (C) mixtures of monomers with complimentary motifs can undergo supramolecular polymerization.

From eqn (4.1) it can be calculated that at millimolar concentrations an appreciable DP for the supramolecular polymer is only achieved when the hydrogen bonding end groups have extremely high binding constants ($<10^6$ M^{-1}). Indeed, the requirement for high bonding constants to deliver high solution state DPs was a driving force behind the production of a range of the self-complementary hydrogen bonding systems by Meijer, Zimmerman and Lehn (Figure 4.2).

The K_a of a specific supramolecular interaction is also dependent on the environment. Consequently, K_a can be rationally and reversibly controlled by changing the conditions of the solution, for example, raising the temperature, altering the solvent or pH of the solution. This leads to a method of changing the DP of the supramolecular polymer, and therefore, the properties of the solution, for example viscosity. Since only the interactions between the supramolecular receptors have been targeted and the covalent bonds are unaffected, subsequent restorations of the system to its original parameters (for example, returning the temperature to the initial value) restores the original DP of the supramolecular polymer and thus the original properties of the solution.

However, the majority of real-world polymeric products are solid materials, where this connection between DP and K_a is not so well-defined. Specifically, factors including crystallinity and phase separation play significant roles in determining the mechanical properties of the bulk polymer. However, the general concept of self-assembly of small, oligomer building blocks into tough, pseudo high molecular weight systems with robust mechanical properties holds true. Moreover, the lower bond energies of the supramolecular bonds when compared to covalent bonds ensures that fractures propagate through scission of the weaker supramolecular bonds rather than covalent linkages. Thus, reforming of supramolecular bonds across fractures results in healing, and this conceptual model applies for the design and realization of the broad range of supramolecular healable polymers that have been studied to date.

The following sections use selected examples from the literature to exemplify the properties of supramolecular healable polymeric materials that are based upon a range of distinct supramolecular interactions.

4.2 Healable Supramolecular Polymers with Hydrogen Bonding Supramolecular Motifs

An early example of harnessing supramolecular polymers to produce healable materials was proposed by Bosman, Sijbesma and Meijer in 2004 using their UPy system (Figure 4.2(A) for the structure of this hydrogen bond receptor). UPy residues can readily be introduced to the terminal positions of many commercially available telechelic polymers (Figure 4.5(A)), including polysiloxanes[25] and poly(ethylene-*co*-butylene) (pE-*co*-B), polyethers, polyesters, and polycarbonates.[26] The impact of the addition of the hydrogen bonding group onto the polymer chain ends can be readily observed as it converts a free flowing liquid starting material (*e.g.* pE-*co*-B) into an elastomeric solid (Figure 4.5(B)). The easy synthesis of these materials from readily available starting materials has been carried out on an industrial scale, resulting in the product Suprapolix™. This material was shown to visibly heal cracks on its surface when heated to 140 °C (Figure 4.5(C)).[22]

It is perhaps not surprising that the addition of highly ordered hydrogen bonding end groups with large K_a values (*ca.* 10^6 M^{-1}) results in such dramatic changes in the physical properties of the solids. However, these systems have the drawback of requiring significant thermal stimulus to enable molecular motion and re-engagement of supramolecular interactions. Therefore, significant effort has focused on studying systems which can exhibit similar healing characteristics but with supramolecular residues that require less energy to promote healing.

4.2.1 Supramolecular Polymers with Low Association Constants

Early studies to investigate the introduction of residues capable of weak hydrogen bonding were conducted in the 1980s and 1990s by the Stadler group.[23] They introduced so-called 'sticky' benzoic acid modified urazole groups (Figure 4.6) onto apolar polybutadiene polymers. This motif can produce a two-dimensional hydrogen-bonded array through conventional donor/acceptor hydrogen bonding on one plane, with these planes assembled into three-dimensional arrays *via* dipole–dipole interactions (Figure 4.6).

Copolymers of urazole monomer (Figure 4.6) with butadiene were found to be thermoplastic elastomers even at extremely low urazole loading levels (>4 mol%).[23c] Stadler proposed the system was adopting a phase separated morphology where the hydrogen bonding modified urazole units were aggregating into discrete "hard" domains, embedded within larger (*ca.* 96%), soft domains of polybutadiene (Figure 4.7). The viscoelastic transition for the self-assembled material determined by rheology was essentially identical to the melting point of hard segments established by DSC analysis. This study demonstrated that the thermoresponsiveness of the material was governed entirely by the weak hydrogen–bonding interactions.

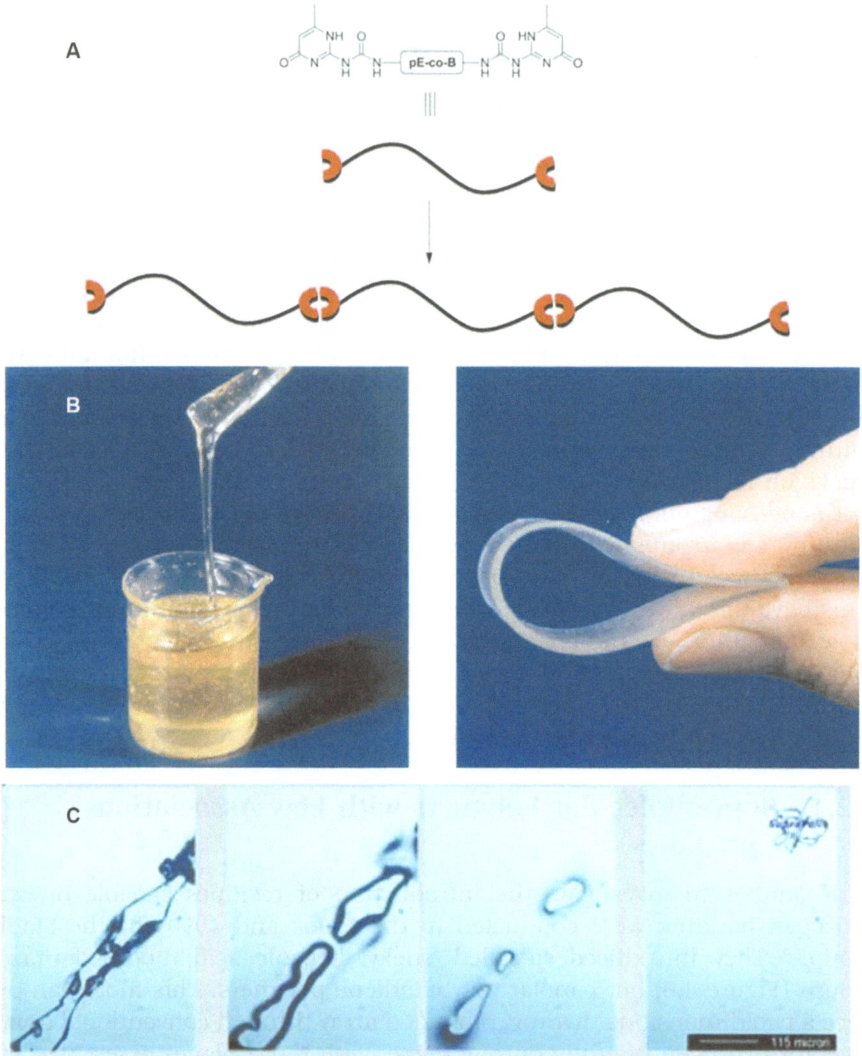

Figure 4.5 (A) Structure and schematic of the UPy functionalized pE-*co*-B healable polymer produced by Bosman, Sijbesma and Meijer; (B) photographs showing the fluid nature of pE-*co*-B and the film forming properties after UPy functionalization; (C) photographs of crack healing at 140 °C. Adapted from ref. 22 with permission from Elsevier, Copyright 2004.

Similar effects caused by the addition of weak hydrogen bonding residues to commercial polymers were observed by Lillya and co-workers.[24] The group modified a low molecular weight poly(tetrahydrofuran) (pTHF, $M_n \approx 2000$ g mol^{-1}, $n \approx 28$) by the addition of terephthalic acid residues to generate two novel polymers that contained either a free acid group

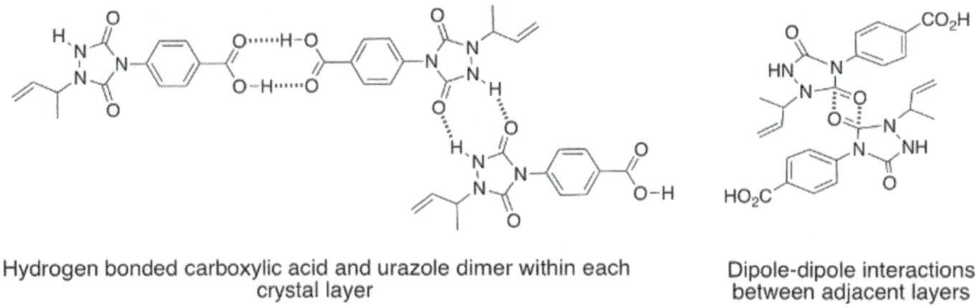

Figure 4.6 Supramolecular interactions (blue dashes) exhibited in the solid state by the urazole residue (A) developed by Stadler and co-workers.

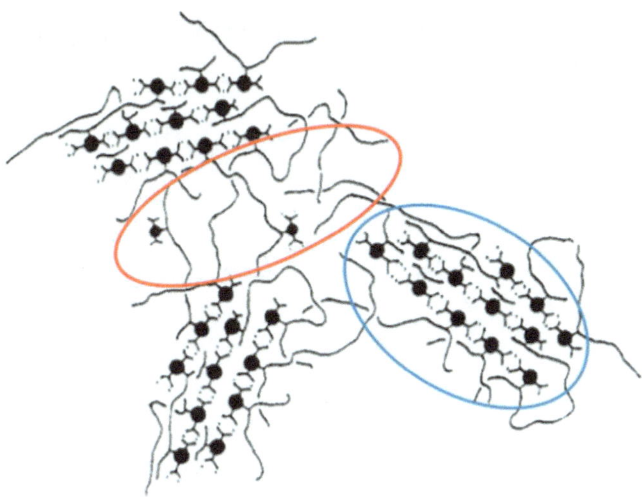

Figure 4.7 Schematic nanoscale structure of the urazole functionalized poly(butadiene) produced by Stadler and co-workers. Urazole-rich hard domains and poly(butadiene)-rich soft domains are highlighted by the blue and red ellipses, respectively. Adapted from ref. 23C with permission from Elsevier, Copyright 1990.

(Figure 4.8(A)) or a benzyl protected acid residue (Figure 4.8(B)). Although both materials were thermoresponsive, the acid terminated system exhibited a storage modulus that was over two orders of magnitude greater than the benzyl terminated system, despite their very close structural relationship. In the solid state the system was found to be phase separated, in common with the materials described by Stadler and co-workers,[23] and contained hard domains consisting of a hydrogen bonding end group with soft pTHF regions lying between.

More recently, Rowan and co-workers produced a series of related supramolecular polymers based upon a central oligomer bis-(3-aminopropyl) terminated pTHF ($M_n \approx 1400\ \mathrm{g\,mol^{-1}}$).[25] This soft and waxy starting material

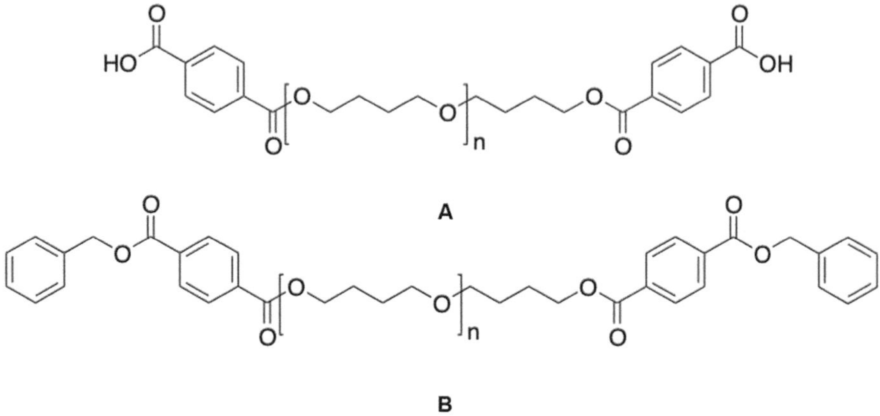

Figure 4.8 Structure of the end-functionalized pTHF (A & B) polymers produced by Lillya and co-workers.[24]

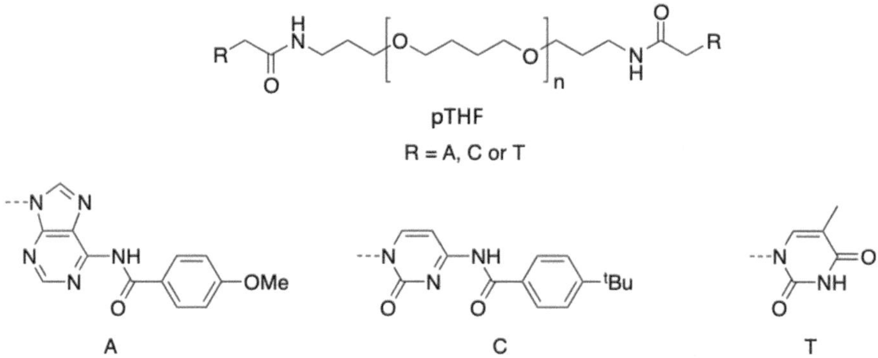

Figure 4.9 Structures of the thermo-responsive nucleobase end-capped pTHF studied by Rowan and co-workers.[25]

was end-functionalized with one of three nucleobases: thymine (T), adenine (A), or cytosine (C) residues (pTHF-A, C or T in Figure 4.9). ^{1}H NMR spectroscopic titration experiments demonstrated that the K_a for each of the nucleobases was lower than 5 M^{-1}, below that required to produce supramolecular polymers in solution with appreciable DP values.[26] After addition of the thymine group to the pTHF, the resulting structure (pTHF-T) was a high-melting solid which could not be cast into free standing films. In contrast, the pTHF modified with adenine (pTHF-A) or cytosine (pTHF-C) could both be solution cast into transparent, free standing films with melting points over 125 °C. However, pTHF-A and pTHF-C also exhibited very disparate properties in the solid state, with the cytosine-terminated polymer presenting as a flexible film which was robust enough to be handled, whereas the adenine-functionalized pTHF-C analog proved to be a brittle material which fractured easily.

The similarity in the structure of the polymers, which only differed by the nature of the end groups demonstrates that, in contrast to solution studies, in the solid state the association constant is just one of a range of factors in determining the properties of solid supramolecular materials. Properties such as phase separation and crystallinity have significant impacts on the final property of the polymer. These observations have been demonstrated to be quite general phenomena and utilized in the design, synthesis and analysis of healable supramolecular materials.

4.3 Healable Supramolecular Polymers Containing Hydrogen Bonding Motifs

An early example of the potential for weakly hydrogen bonded polymer systems to have useful healing properties was disclosed by Chino and Ashiura in 2001 arising from their studies to produce recyclable alternatives to vulcanized rubber.[27] They postulated that replacement of the covalent bonds in thermoset rubber networks with reversible non-covalent hydrogen bonding groups would furnish a product that could be thermally re-formed. The group synthesized a polyisoprene containing up to approximately a 6 mol% loading of triazole residues (Figure 4.10). This procedure transformed the based isoprene polymer from a fluid to a thermoresponsive solid. Interestingly, this material did not exhibit creep phenomenon, observed typically in thermoplastic materials, indicating that the supramolecular bonds were sufficiently immobile under the experimental conditions used, making the material behave more akin to a crosslinked thermoset. However, in contrast to crosslinked materials, it could be re-formed up to 10 times without changing its mechanical properties. Whilst not 'healing' in the traditional sense (*e.g.* removal of a crack) this study demonstrated the concept of harnessing the reversibility of supramolecular hydrogen bonding interactions to produce materials that could be fractured and reformed without loss of strength.

In 2008 Leibler and co-workers produced the first weakly hydrogen bonded material that was designed and experimentally verified to undergo several break-and-heal cycles.[28] The structurally diverse polymer was generated from vegetable fatty acid derivatives, diethylene triamine and urea. The resulting oligomeric species were polydisperse in nature containing up

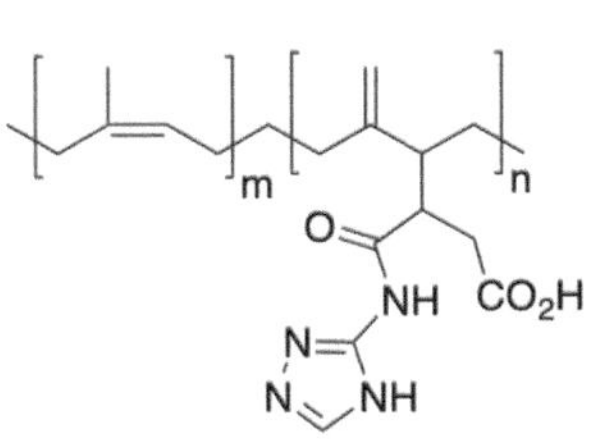

Figure 4.10 The structure of a recyclable supramolecular polymer reported by Chino and Ashiura (m–n = 100; n < 5.7).[27]

to around 7–9 complimentary and self-complimentary hydrogen bonding groups (Figure 4.11).[29] The initial product was a glassy material at room temperature ($T_g = 28$ °C), although this was transformed into an elastomer with a T_g of 8 °C and an elongation at break of 600% by the addition of 11 wt% dodecane. Break/heal experiments were conducted by cutting a sample of the plasticized supramolecular polymer with a razor and then pressing the two fractured surfaces together by hand. The rejoined, healed material was found to exhibit essentially identical properties to the pristine (un-cut) sample. However, it was found that when the cut samples were left separated for several hours before applying the pressure induced healing procedure a material was afforded whose properties were diminished in comparison to the pristine sample. This reduction in healing efficiency was attributed to the mobility of the chains at the surface allowing the freshly revealed hydrogen bonding residues to defuse back into the bulk, and, therefore, not available to reform non-covalent networks across the fracture site.

In 2012, Guan and co-workers designed an autonomously healable supramolecular system using multiple weak hydrogen bonding interactions.[30] A brush type polymer was synthesized which consisted of an apolar polystyrene backbone (DP≈114), each chain containing around 11 highly polar polyacrylic acid-amide side chains (DP≈186, see Figure 4.12(A)).

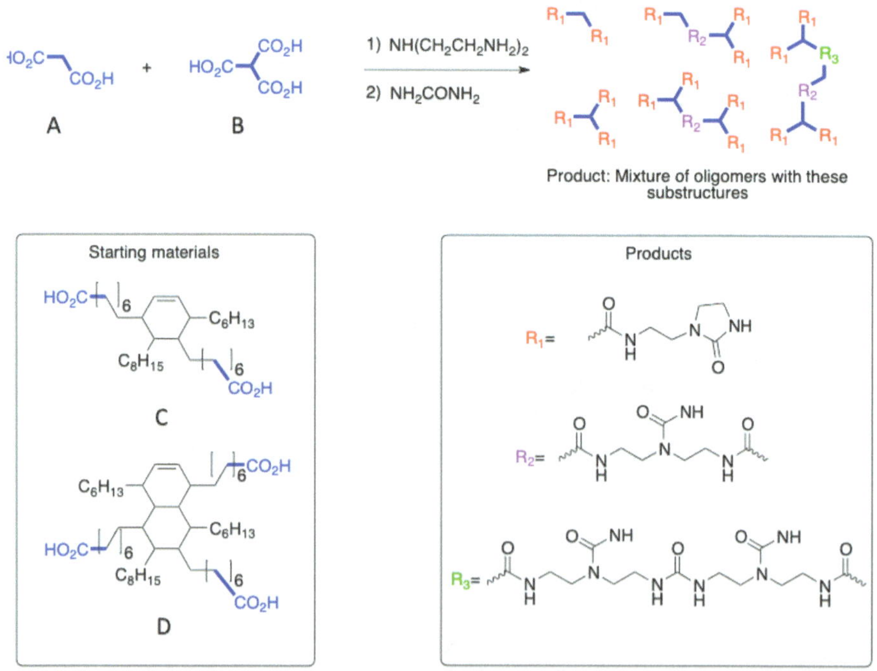

Figure 4.11 Synthesis of the branched polymer produced by Leibler and co-workers that exhibited self-healing properties after swelling in dodecane.

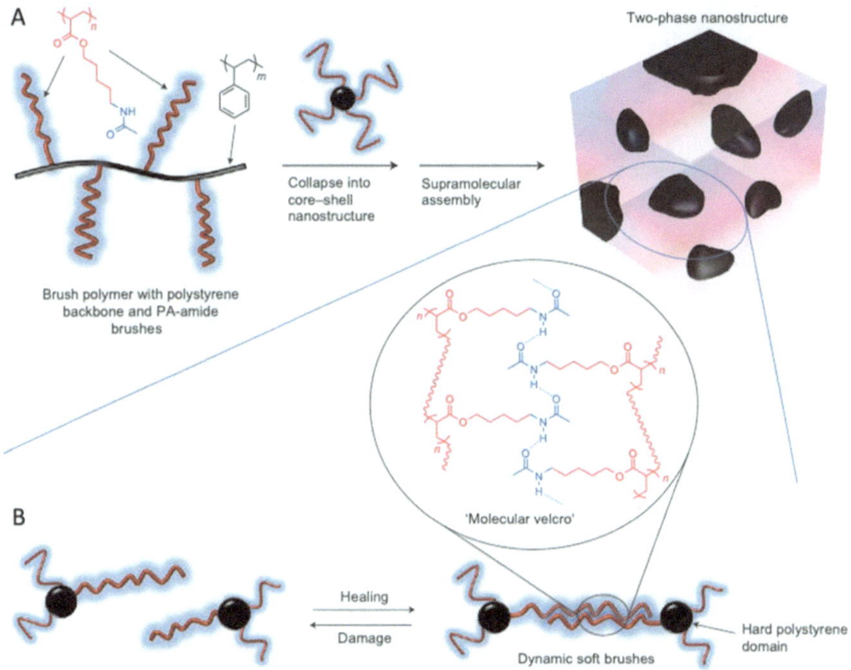

Figure 4.12 (A) Structure and schematic portrayal of the assembly of the supramolecular, healable phase-separated brush co-polymer produced by Guan and co-workers. (B) Schematic of the proposed rupture and reforming of the supramolecular bonds formed between hydrogen bonding residues of the polymer brushes during a break/heal cycle. Adapted from ref. 30 with permission from Springer Nature, Copyright 2012.

The immiscibility of the apolar backbone and polar side groups resulted in the system adopting a core–shell type structure. This elastomeric material was found to be nano-structured (Figure 4.12(B)), containing discrete polystyrene-rich hard domains connected by a continuous phase which was rich in hydrogen bonded polyacrylic acid side chains ($T_g \approx 5$ °C). Break/heal type experiments revealed healing efficiencies of over 90% after a healing time of 24 hours at room temperature.

The weakly hydrogen bonded healable supramolecular systems discussed above have highly designed end groups or relatively complex and dispersed macromolecular architectures. In 2018, the Aida group produced a linear polythiourea (pTU, Figure 4.13(A)) system which did not contain complex end groups or branching.[31] The pTU was amorphous, mechanically robust and non-tacky yet was able to self-heal at freshly cut surfaces by compression within 30 minutes. An analogous polyurethane (PU, Figure 4.13(B)) produced for direct comparison was semicrystalline in nature and exhibited a tensile modulus value approximately 10% of the value exhibited by pTU (0.2 GPa *vs.* 1.9 GPa). The considerable difference in physical properties between the two

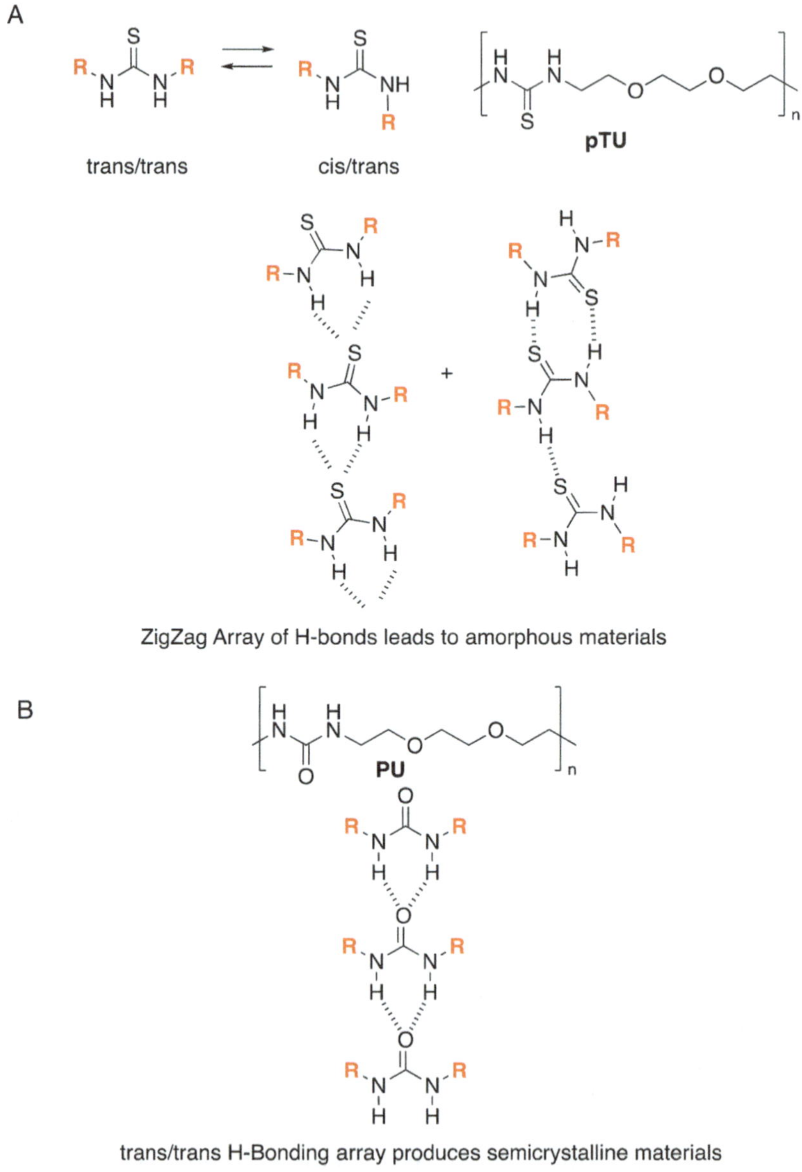

Figure 4.13 (A) Structures of the *cis* and *trans* isomers possible in thiourea systems and the zigzag arrangement of hydrogen bonding arrays that leads to an amorphous structure in thiourea systems; (B) linear hydrogen bonding arrays that led to a semicrystalline structure in the analogous urea-based polymer.[31]

structures was attributed to the 'zigzag' nature of the hydrogen bonding in the pTU when compared to the linear hydrogen bonding in the PU, the regularity of which produced the observed semicrystalline material.

4.4 Healable Supramolecular Polymers Utilizing π–π Stacking Interactions

The underlying design criteria for producing a supramolecular healable material is that the dynamics of the supramolecular interactions should be sufficiently rapid to allow healing under appropriate stimuli and within useful time periods. Therefore, in principle any supramolecular interaction that meets these criteria could be used to produce a healable material. Indeed, supramolecular interactions that do not use hydrogen bonding may offer additional benefits, for example, their strength is less likely to be diminished by absorption of atmospheric moisture which can serve to weaken the hydrogen-bonded arrays.

The interaction between electronically complementary aromatic π-systems has formed the basis of the synthesis of an extraordinarily diverse range of mechanically interlocked molecules (MIMs).[32] Specifically, the reversible association π-electron rich and π-electron poor systems[33] has underpinned many of the key developments on switchable materials from the Stoddart laboratory which ultimately contributed to the award of the 2016 Nobel Prize in conjunction with Sauvage and Feringa.[34]

With respect to the field of healable materials, Colquhoun, Hayes and co-workers incorporated π–π stacking interactions into a range of well-defined supramolecular polymer blends which were able to undergo repeatable crack healing in response to small increases in temperature. The major design element in their approach to healable systems was a 'chain-folding' structure,[35] which comprised two π-electron deficient naphthalene-1,4,5,8-tetracarboxylic diimide units separated by a flexible diether linking unit (Figure 4.14(A)). Computational modeling confirmed by experimental studies on synthesized systems showed that as the number of face-to-face π–π-stacking interactions increased from two to four (Complexes A to C) the binding constant increased by two orders of magnitude from 130 M^{-1} to 11 000 M^{-1}.[36]

The chain folding diimide unit could be introduced into the main chain of a range of polymeric systems, with pyrene terminated polymers being accessed by functionalization of commercial telechelic polymers such as PDMS (Figure 4.15(A) and (B)).[37] Blends of these two polymers were found to be relatively weak materials but were visually homogeneous and underwent crack healing in response to a thermal stimulus.

By maintaining the π-electron deficient and π-electron rich residues that are present in each component of the blend, the main chains of the polymers could be structurally altered to produce a second generation of materials with improved physical properties (Figure 4.15(C) and (D)). This was achieved by introducing an oligomeric diamine (Jeffamine®) into the system to produce two new polymers, that when blended could be cast into more robust free-standing films.[38]

Crack healing of the polymer blend produced from the generation two polymers (Figure 4.15(C) and (D)) was visualized by variable temperature

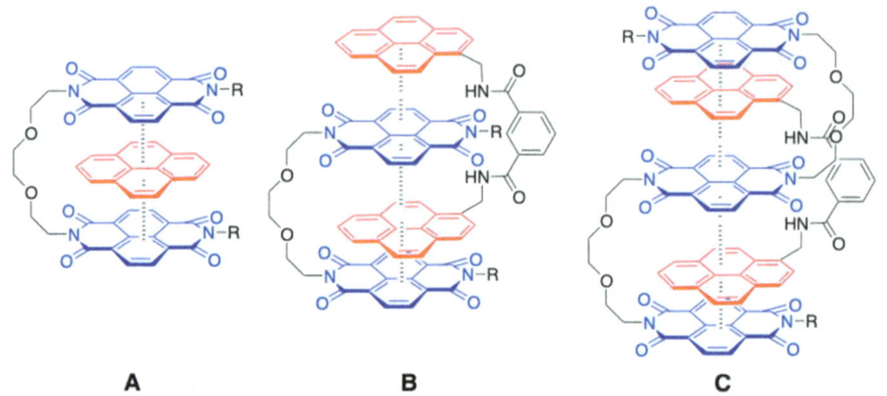

Figure 4.14 Structures of the complexes designed and modeled (R = Me) to produce a series of related supramolecular assembles supported by π–π stacking interactions. The complexes (R = 2-ethyl hexane) were synthetically verified to have binding constants of 130, 3500 and 11 000 M^{-1} (A to C, respectively).

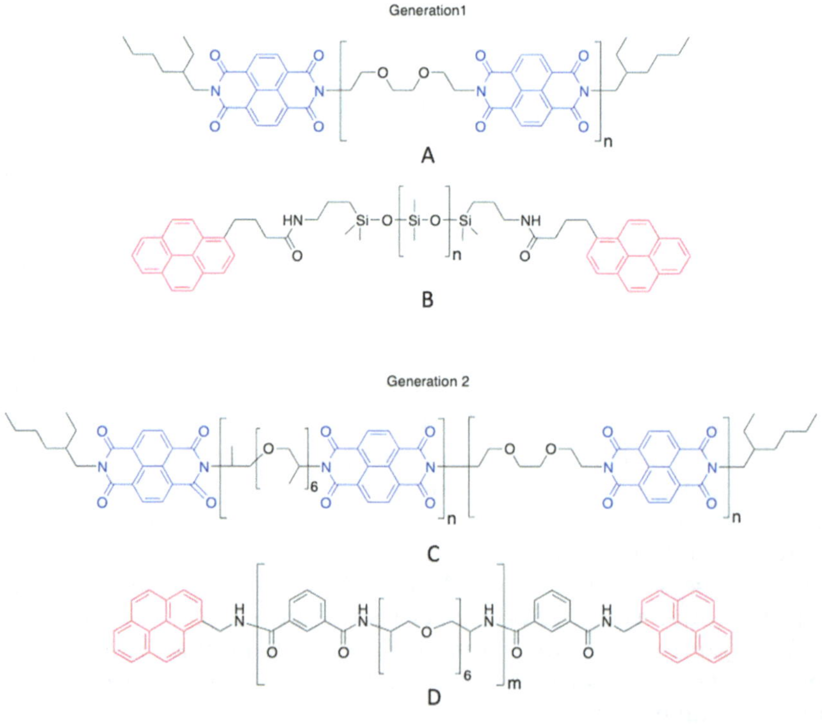

Figure 4.15 The structures of two generations of healable supramolecular polymers produced by Colquhoun, Hayes and co-workers.

environmental scanning electron microscopy (ESEM). Through this technique, fractures of approximately 70 μm were rapidly (less than a minute) eradicated by heating the material above 80 °C (Figure 4.16).

A more quantitative investigation into the healing properties of the generation 2 material was carried out by stress strain analysis of samples that had undergone cycles of breaking/healing. In this process, samples were cut in two, before the edges of the cut sample were overlapped and then placed in contact for varying periods of time (up to 10 minutes) at 50 °C. The material exhibited a healing efficiency (η_{eff}) of 100% after approximately 5 minutes healing time (Figure 4.17). In addition, the same sample was able to undergo 3 break–heal cycles with approximately 100% η_{eff}. In contrast to several supramolecular polymers with healing characteristics that were based on hydrogen bonding assemblies, in this case recovery of the pristine properties was also observed after the edges of the broken sample were separated for 24 hours.

Further architectural variants of polymer blends with these π-electron rich and poor structural motifs were also produced and tested. Specifically, polymers that contained either dipyrene tweezer type end-groups (Figure 4.18(A)) or hydrogen bonding residues in the main chain of one of the polymers (Figure 4.18(B)) were produced to compare the impact of these structural variations in a systematic way.[39]

The addition of dipyrenyl end groups to the polymer was predicted to enable the formation of complexes akin to B and C shown in Figure 4.14, which have significantly higher binding constants than those possible when only one pyrene end group is present (Figure 4.14(A)). Indeed, when introduced into polymeric systems the dipyrenyl end-capped polymers produced a significant enhancement in the physical properties in the resulting materials. Blends of dipyrene tweezer end-capped polymer (Figure 4.18(A)) and

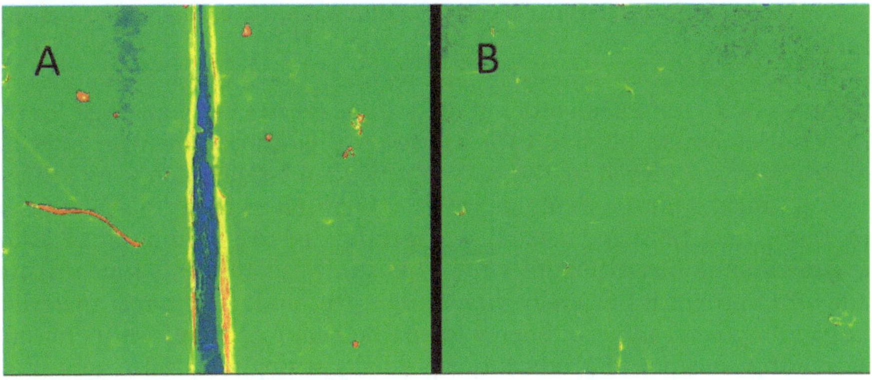

Figure 4.16 ESEM images of a fracture zone of the supramolecular healable at ambient temperature and 87 °C (images A and B, respectively). The fracture is approximately 70 μm across. Reproduced from ref. 38 with permission from the Royal Society of Chemistry.

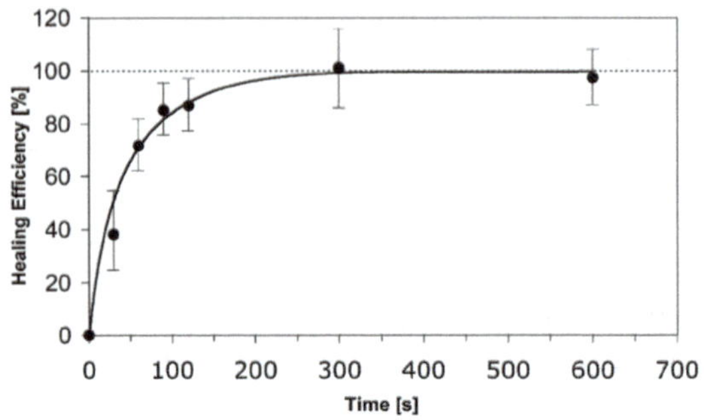

Figure 4.17 Plot of healing efficiency (and tensile modulus) as a function of healing time at 50 °C for the 2nd generation supramolecular healable blend produced by Colquhoun, Hayes and co-workers. Reproduced from ref. 38 with permission from the Royal Society of Chemistry.

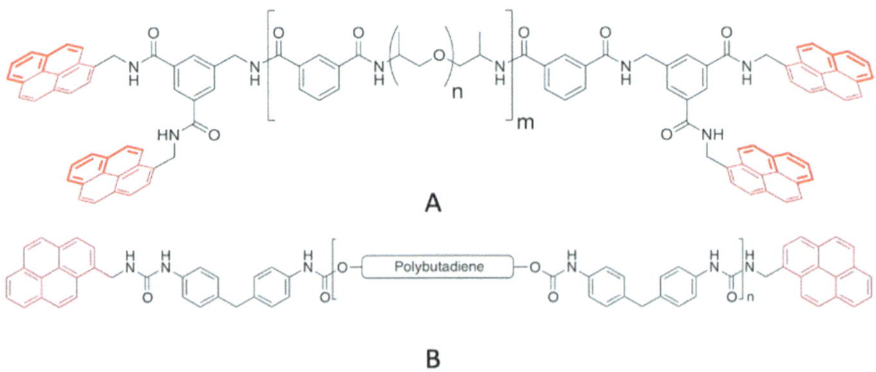

Figure 4.18 Structures of the dipyrene end-capped polymer (A) and pyrenyl end-capped polyurethane (B) used in healable two component supramolecular polymer blends by Colquhoun, Hayes and co-workers.

chain folding polymer shown in Figure 4.15(C) exhibited a modulus of toughness of 3×10^8 Pa, an order of magnitude greater than observed for the generation 2 system (Figure 4.15). However, this improvement in physical properties was achieved at the cost of extended healing times to achieve 100% η_{eff} for the tensile modulus which was observed to be 160 minutes at 140 °C (Figure 4.18(A) and 4.15(C)) when compared to 5 minutes at 50 °C for the generation 2 blend (Figure 4.15). This study exemplifies a fundamental trade-off in properties whereby increasing K_a to produce stronger materials results in a greater input of energy to enable healing. Therefore, increasing the binding constant to produce exceptionally strong materials could push the healing conditions into the region where the material undergoes thermal decomposition.

In a distinct set of studies, pyrenyl end-capped PU was produced (Figure 4.18(B)), through a co-polymerization of methyldiphenyldiisocyanate with dihydroxypolybutadiene.[40] This simple, one-pot synthesis produced a polymer containing hydrogen bonding urethane and urea residues in the backbone whilst retaining the pyrenyl end groups required to form π-stacked complexes with NDI-containing chain folding polymers such as that shown in Figure 4.14(A). Blends of this hydrogen bonding polymer (Figure 4.18(B)) and chain folding polymer (Figure 4.15(C)) were tough, elastomeric products which could be extended up to 170% prior to fracture, when compared to extensions at break values of just 70% elongation for the generation 2 material (C and D in Figure 4.15 – this blend lacks extensive hydrogen bonding). Despite the addition of hydrogen bonding to re-enforce the original π–π stacking interactions, this new polymer blend showed excellent healing properties and retained 91% of the original elasticity and 70% of the modulus of toughness when compared to those values exhibited by pristine samples.

In 2014, Hayes and Colquhoun reported the effect of cross-linking 'density' on the properties of π–π stacked supramolecular healable materials. Blends featuring a chain-folding NDI containing polymer (Figure 4.19(B)) and either a linear polyether (PE) equipped with two pyrene end groups (Figure 4.19(A)), or a tri-armed PE (Figure 4.19(C)) end-capped with three pyrenyl groups were produced.[41] Comparison of the blends featuring the two and three-armed systems revealed an increase of approximately 60% in tensile modulus upon increasing the cross-link density. The enhanced tensile modulus was also retained over 3 break–heal cycles under comparable thermal healing conditions. More dramatically, the modulus of toughness for the blend featuring the tri-armed PE system was found to be an order of magnitude greater (1.45 MPa *vs.* 22.6 MPa) when compared to the network with the linear polymer PE.

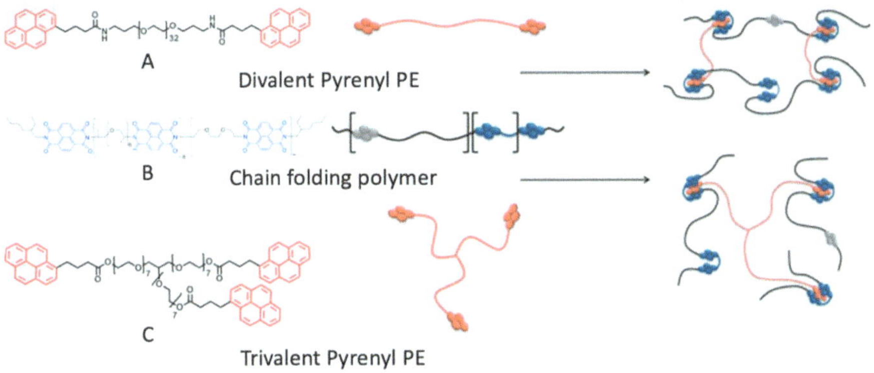

Figure 4.19 Structures of the polymers synthesized by Hayes and co-workers studying the effects of cross-linking 'density' on healable supramolecular blends together with schematics of the structures of the blends. Adapted from ref. 41 with permission from the Royal Society of Chemistry.

This section has summarized research concerning healable materials that use π–π stacking supramolecular interactions to produce healable polymeric systems. Through iterative optimization processes, these polymers clearly demonstrate the underlying design criteria than can be used to optimize a range of related supramolecular polymers, specifically: (i) that increasing the association constant within the material leads to tougher materials, although at the cost of requiring greater energy input to facilitate healing; (ii) maintaining the association constant but increasing the cross-linking density enhances both tensile modulus and modulus of toughness without dramatically altering the healing conditions; and (iii) orthogonal supramolecular interactions (namely π–π stacking and hydrogen bonding in this case) can be built into the design of the materials to produce polymers with enhanced physical properties.

4.5 Healable Supramolecular Polymers Utilizing Halogen-bonding Interactions

Although a less well studied form of supramolecular interaction, systems that contain halogen bonding were produced over two centuries ago.[42] However, they have only been intensively studied over the last few decades. During this latter period, structures containing designed halogen bonding motifs have become increasing widely studied with many potential applications proposed and realised.[43]

Halogen bonding arises from the interaction between specific classes of Lewis acids and Lewis bases which behave as electron donor and acceptor species, respectively, in the supramolecular bond. Whilst Lewis acid/base pairs are well-known in chemistry, in the case of halogen bonding, the donor component is a halogen atom which is covalently bonded to an R group that is itself highly electronegative. This structural arrangement results in the strong polarization of the R-halogen bond which can lead to the formation of a (partial) positive region on the far side of the R-halogen bond, which is termed the σ-hole (see the red colored region of the structure shown in Figure 4.20). Recently the sigma-hole has been observed experimentally for the first time.[44]

The presence of the sigma-hole allows these structures to behave as an electrophile, accepting electron density from Lewis bases. As a consequence, halogen bonding interactions are highly tunable both in terms of strength and directionality. Typically, strength increases with the polarizability of the halogen (I > Br > Cl > F) and the polarizing strength of the R group to which the halogen is attached. As the position of the sigma hole halogen can be precisely calculated, the angle and strength of the supramolecular bond produced with differing nucleophilic species can be readily approximated.

With respect to polymeric systems (though not necessarily healing systems), some of the most widely studied halogen bonding residues are shown in Figure 4.21.

Figure 4.20 Schematic of the surface potential of CF_3I with the sigma hole shown in red. Adapted from ref. 45 with permission from Springer Nature, Copyright 2006.

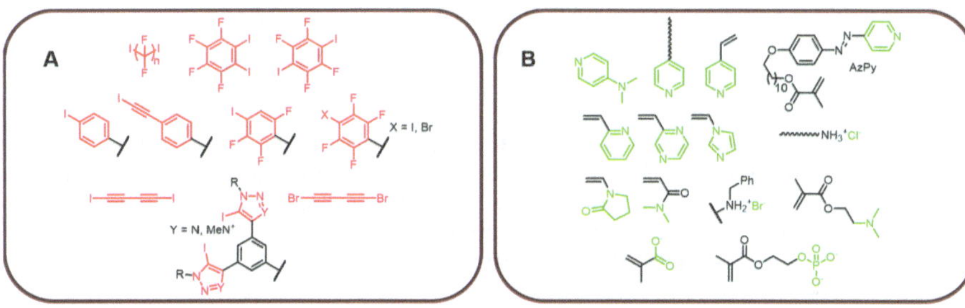

Figure 4.21 Structures of halogen bonding donor and acceptor groups (A and B) that have been used in supramolecular polymer chemistry. Adapted from ref. 46 with permission from the Royal Society of Chemistry.

Studies in the area of healing polymer systems containing halogen bonding were initiated by Schubert and co-workers.[47] The team produced an organic phosphate containing ionomer (P2, Figure 4.22) which was cross-linked through halogen bonding interactions by a bis-bidentate iodine containing linker (Linker A, Figure 4.22) to produce the supramolecularly crosslinked system (P2A in Figure 4.22).[47b] Comparative nanoindentation studies with model ionomers and linkers that lacked the halogen bonding residues showed that the strength of the base ionomer was enhanced by a factor of 10 in the presence of the halogen bonding cross-linker. Crack healing was observed in the halogen bonded system when heated to 100 °C for 3.5 hours (Figure 4.22 photos A and B).

4.6 Healable Polymeric Materials Containing Inclusion Complexes

Inclusion complexes generally comprise of a macrocycle that has a cavity with suitable dimensions to encapsulate a second chemical entity. Frequently the driving force for complexation is provided by van der Waals forces. However,

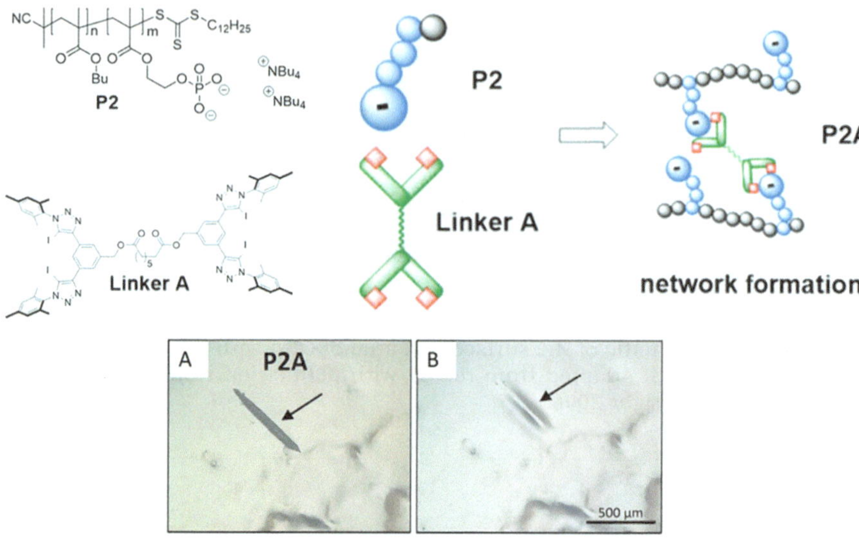

Figure 4.22 Structures and schematic of the formation of the halogen bonded network studied by Schubert *et al.* and optical microscope images of the (a) scratched surface and (b) surface after thermal healing at 100 °C for 3.5 hours. Adapted from ref. 47b with permission from the Royal Society of Chemistry.

depending on the chemical structures in question, additional interactions (for example, hydrogen bonding or dipole–dipole interactions) may also enhance complexation. There are a range of well-studied macrocycles suitable for forming inclusion complexes, for example, cucurbiturils, cyclodextrins or calixarenes.

There have been significant investigations on the use of inclusion complexes within healable gels, which is discussed in detail in a later section (Section 4.8). In recent years several solvent-free variants of these have been produced, as exemplified by the studies of Harada and co-workers who made use of the reversible nature of the inclusion complex between γ-cyclodextrin and norbornene.[48] The polymer was produced by co-polymerization of acrylate functionalized monomers of γ-cyclodextrin and norbornene with unfunctionalized ethylacrylate (see pEA-PAcyCD-Ad in Figure 4.23). The resulting elastomeric material exhibited a healing efficiency of essentially 100% after cut edges were placed in contact for 10 hours at 80 °C. In stark contrast, control materials (*e.g.* pEA-BDA in Figure 4.23) that did not feature the monomers with the supramolecular motifs showed minimal healing under the same thermal processing conditions (η_{eff} values less than 20% were obtained).

4.7 Healable Supramolecular Polymers Utilizing Metal–Ligand Interactions

Metal–ligand interactions underpin the field of organometallic chemistry. As such there is an extensive body of literature focused on understanding,

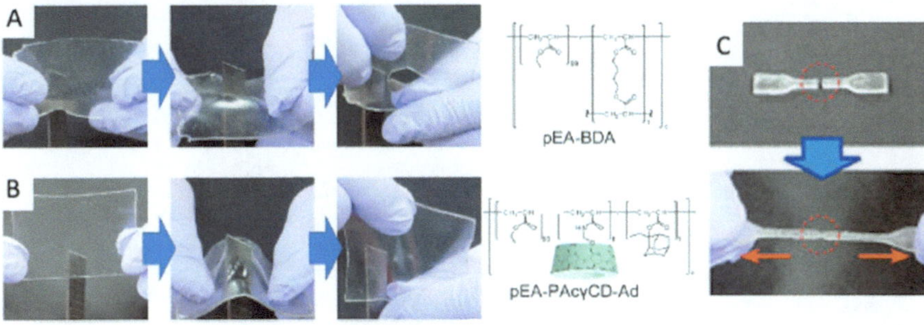

Figure 4.23 (A) Photographs showing the ease of damage of pEA-BDA that does not contain supramolecular cross-links; (B) example of the resistance to breaking of the pEA-PAcyCD-Ad which contains supramolecular cross-links; (C) example of the break/heal test carried out on pEA-PAcyCD-Ad. Adapted from ref. 48 with permission from American Chemical Society, Copyright 2019.

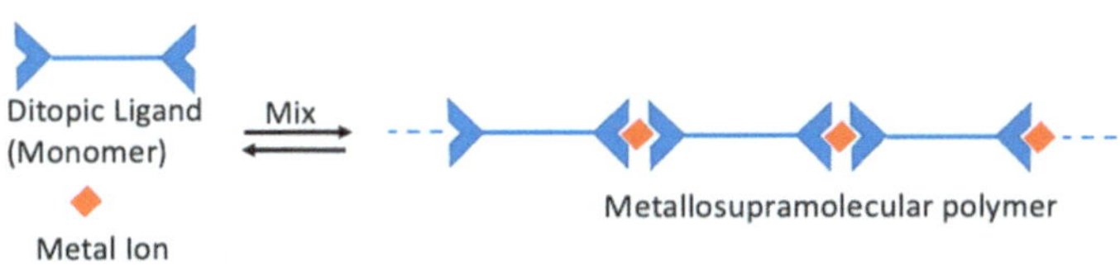

Figure 4.24 Schematic synthesis of a metallosupramolecular polymer from mixing ditopic ligands and metal ions.

prediction and tuning of the strength and directionality of interactions between principally organic based ligands and metal centers. As a consequence of the range of metallic elements in the periodic table, their variance in electronic configuration as well as the range of ligand systems that have been explored, the nature of metal–ligand interactions can vary from essentially covalent to distinctly ionic in nature. The reversible nature of metal–ligand interactions plays a fundamental role in the majority of reactions involving organometallic species, be they biological enzymic systems or entirely synthetic organometallic catalytic species.

Conceptually, adding stoichiometrically appropriate quantities of metals ion to oligomeric or polymeric systems which have been functionalized with appropriate ligands can result in the formation of metallosupramolecular polymers (Figure 4.24).[49] The reversible formation of the metallosupramolecular polymer enables the construction of healable bulk polymer materials.

The first example of a designed synthesis of a metallosupramolecular polymer that was experimentally verified to undergo healing was produced by Rowan, Weder and co-workers.[50] Their system consisted of a low molecular weight, low T_g macromolecule, (pE-*co*-B), whose chain ends were functionalized with 2,6-bis(1′-methylbenzimidazolyl)-pyridine (Mebip) ligands. These ligands are known to form stable complexes with both

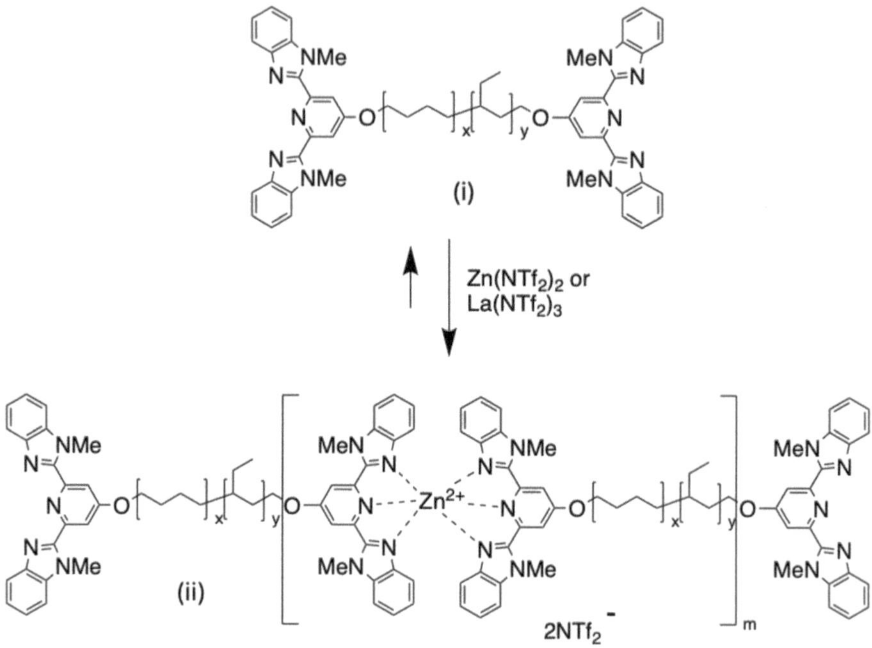

Figure 4.25 Chemical structures of the macromonomer (i) and the healable metallosupramolecular polymer (ii) produced by Weder, Rowan and co-workers.[50]

transition metals (*e.g.* Zn) and a range of lanthanide (Ln) ions (Figure 4.25). The resulting highly elastomeric supramolecular polymers did not have electronic absorption bandings in the visible region of the UV-vis spectrum so were essentially colorless. However, the Mebip ligand and its Mebip/metal complexes did exhibit an absorption band with maxima at *ca.* 310 nm and 340 nm, and, therefore, absorb UV radiation. The low fluorescence quantum yield of the complexes facilitates the energy of incident UV radiation (320–390 nm and an intensity of 950 $mW\,cm^{-2}$) to be converted to heat, which produced a measurable elevation in the temperature of the polymer (up to 200 °C). Since the bulk material exhibited low thermal conductivity, advantageously only the area under irradiation was subject to an increase in temperature. Therefore, it was proposed that this system could be healed *in situ* and even under load without complete rupture of the polymer.

This concept of photo-thermal healing would appear to be applicable to many supramolecular polymeric systems provided they efficiently convert photons to heat. Furthermore, there are several polymeric systems that are known to exhibit a mechanochromic response[51] in that they undergo a color change at the damage site. Combining damage sensing by color change with photon-induced healing may be a suitable method to enable autonomous self-healing of materials whereby light is only absorbed at defect sites.

4.8 Introduction to Supramolecular Gels

Supramolecular gel materials have been studied extensively in the past 25 years. They have seen particularly intense research with the aim of developing them for applications as diverse as biomedical science and electronic devices.[52] The general structure of these materials is exemplified in Figure 4.26 which shows a gelator network which sequesters the solvent, be that water (a hydrogel) or organic solvent (an organogel). As evident from the schematic in Figure 4.26, the gelator to solvent ratio is low, typically between 0.5 and 10 wt% solids.

Supramolecular gels can be further subdivided into polymeric and molecular gelators. In the former case, the gel is maintained by macromolecules which self-assemble to sequester the solvent, whereas in the latter case, small molecules assemble into supramolecular polymers, which undergo a further level of organization to produce the interconnected network which ultimately yields the gel its physical properties. In both polymeric and supramolecular gel systems, the high solvent ratio results in generally weak materials (G' less than 500 Pa, see Chapter 1) which preclude their use as load bearing structural materials. However, there are a growing range of applications including battery technology or tissue engineering where this relative lack physical robustness is not a disadvantage and the ability to heal defects would be extremely beneficial.

The types of supramolecular interactions found in gels are the same as those discussed in previously sections (*e.g.* hydrogen bonding, π–π stacking, metal ligand interactions, *etc.*). Fractures occur through disengagement of these interactions and subsequent healing is achieved through re-engagement of the supramolecular bonds across the fracture interface. To achieve efficient re-engagement requires rapid molecular motion, which is facilitated in 'bulk' (non-gel) materials when the materials are 'healed' above the T_g of the system. In gels, fast molecular dynamics are ensured by the

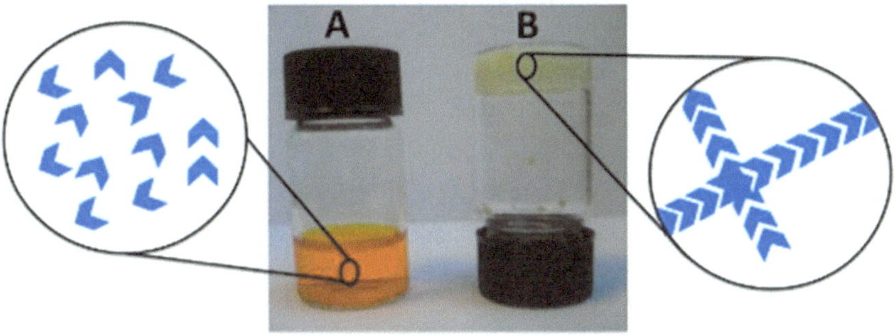

Figure 4.26 Photographs of a supramolecular gelator in solution (vial A) and gel state during a tube inversion test (vial B). The schematics show the supramolecular gelator (blue arrows) in solution and after self-assembly (left and right, respectively).

presence of the solvent, which facilitates rapid healing as demonstrated in the selected examples that are presented in the following sections.

4.8.1 Healable Supramolecular Polymeric Gels

In 2011 Harada and co-workers produced an interesting example of a redox healable polymeric hydrogel gel by exploiting the host–guest interactions between ferrocene and β-cyclodextrin.[53] In its native state, ferrocene is known to form a strongly-bound inclusion complex with β-cyclodextrin, whereas oxidation to produce the ferrocenyl cation results in decomplexation of the system (Figure 4.27).

The healable material comprised of two structurally related polymers based on a poly(acrylic acid) ($M_w = 25$ kDa) backbone. One polymer contained approximately 5 mol% β-cyclodextrin, whereas the second component was substituted with approximately 2.7 mol% ferrocenyl residues. A mixture of these two polymers produced a stable hydrogel at a concentration of 2 wt%. Chemical (NaClO) or electrically induced oxidation of the ferrocenyl side groups resulted in a loss of integrity of the gel as a consequence of the reduction in strength of the host–guest complex. The dissociation was reversed by reduction of the ferrocenyl cation back to the neutral form (either chemically or electronically).

Healing experiments were conducted by cutting the hydrogel and placing the freshly-cut edges in contact for 24 hours at room temperature, over which time period the two sections fused to re-form a single hydrogel with approximate 84% healing efficiency (Figure 4.28). Control experiments demonstrated that addition of a chemical oxidant, or a competitive chemical guest for the β-cyclodextrin to the cut surface of the hydrogel resulted in a loss of healing ability of the system as a consequence of the disruption of the host–guest interactions.

In 2012 Phadke *et. al.* produced a lightly cross-linked polymer hydrogel and demonstrated its use for producing healable chemically resistant coatings, tissue engineering scaffolds and in drug delivery applications.[54]

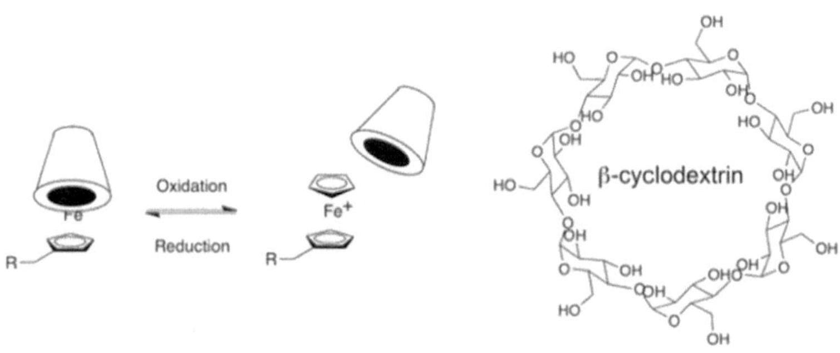

Figure 4.27 Schematic of the redox induced reversible complexation of β-cyclodextrin with ferrocenyl derivatives.

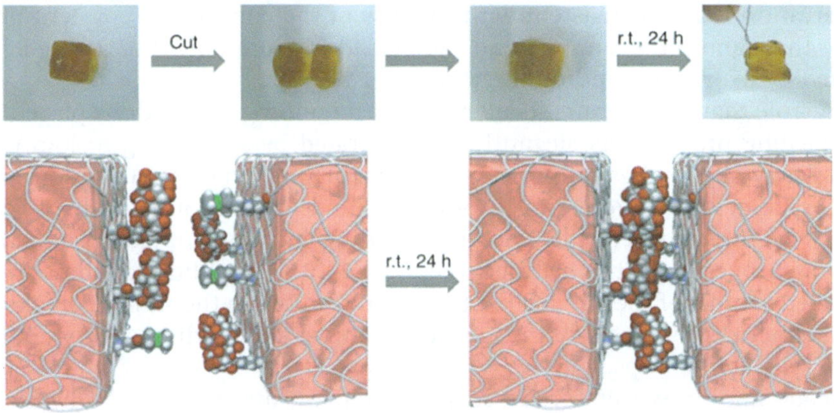

Figure 4.28 Example of a healing experiment where two severed pieces of hydrogel were placed in contact for 24 hours and fused. The schematic shows the proposed reversible engagement of the ferrocenyl (iron atoms in green) and β-cyclodextrin residues during healing. Adapted from ref. 53 with permission from Springer Nature, Copyright 2011.

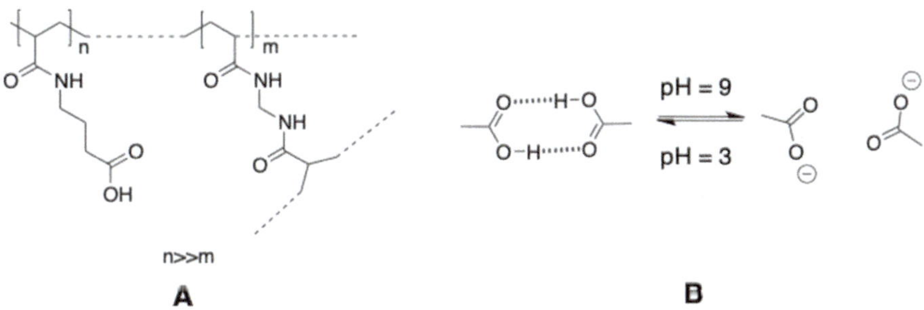

Figure 4.29 (A) Chemical structure of the healable hydrogel possessing both supramolecular and chemical cross-links. (B) Scheme showing the proposed pH switchable supramolecular interactions (blue dashes) required to promote healing.

The gel comprised of a loosely covalently cross-linked acrylonitrile network with multiple carboxylic acid side groups (Figure 4.29(A)). Typically, cross-linked polymeric gels would not exhibit healing properties unless dynamic covalent bonding units are present (see Section 3.3) to establish new covalent linkages and bridge the fractured surfaces. In this example, at pH = 3 the pendent acid side groups are protonated and can form additional supramolecular cross-links through hydrogen bonding interactions. Thus, at low pH placing two pieces of this gel in contact for even a few seconds results in the two sections of gel becoming fused, although increased contact time (24 hours) resulted in stronger joints. The healing efficiency of this system was modest (66%) when compared to other supramolecular systems, an observation that is consistent with the loss of the continuous covalent network throughout the sample, and the healed fracture only then being supported by hydrogen bonding interactions.

The role that the supramolecular hydrogen bonding interactions played in the healing process was elegantly demonstrated by attempting to join two samples swollen in basic aqueous solutions ($pH = 9$). In this pH window, the carboxylic acid residues are deprotonated, and cannot form the hydrogen bonds required to heal the fracture, resulting in the complete loss of the healing properties of the hydrogel (Figure 4.29(B)). In addition, submerging a healed sample of the gel in a urea solution resulted in the sample breaking at the fracture point as a consequence of competitive binding of urea to the carboxylic acid groups.

Huang and co-workers produced a supramolecularly cross-linked hydrogel employing the well-known interactions between quaternary ammonium salts and crown ether residues that form pseudorotaxane assemblies (Figure 4.30(A)).[55] The two component system comprised of a poly(methyl methacrylate) backbone ($M_n = 9.5$ kDa) which was functionalized with dibenzo[24]crown-8 (DB24C8) side groups (approximately 7 per polymer chain) (Figure 4.30(B)). This polymer was soluble in chloroform/acetonitrile (50 : 50 v/v) at room

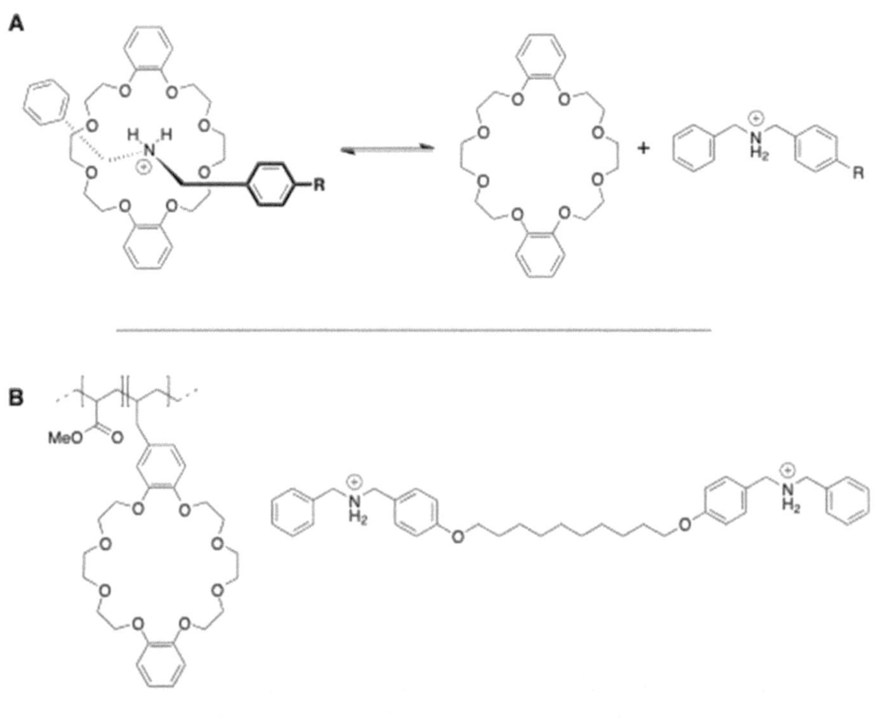

Figure 4.30 (A) Reversible formation of a crown ether/ammonium ion pseudorotaxane. (B) Structures of the crown ether functionalized poly(methyl methacrylate), and ditopic ammonium ion cross-linker used to form supramolecular healable gels by Huang and co-workers.

temperature but formed gels immediately on addition of the ditopic ammonium ion cross-linker (Figure 4.30(B)). Rheological healing tests (see Section 1.5.3) showed that the supramolecular gel could undergo multiple break–heal cycles with healing efficiencies above 95%.

4.8.2 Healable Supramolecular Molecular Gels

Although there are several reports of healable low molecular weight gelators, the mode of healing is frequently through dynamic covalent chemistries, rather than supramolecular interactions,[56,57] and are therefore beyond the scope of this chapter. Recently, however, Dastidar and co-workers[58] have used the supramolecular synthon approach[59] to guide the design of structurally simple organogelators derived from ammonium salts of amino acids derivatives. One of these gels, comprised from Boc-protected glycine (Figure 4.31(A)) and dibenzylamine (Figure 4.31(B)) (7 wt% in nitrobenzene) was observed to form free-standing gels that proved to be stable for many months under ambient conditions. Healing was visually demonstrated by placing five cubes of gel in contact for 1 minute, which resulted in the pieces fusing to form a free-standing cuboid of gel (Figure 4.31(C)). Rheological healing studies (see Section 1.5.3) demonstrated essentially instantaneous healing of the sample after shear fractures.

The select examples described in Section 4.8 demonstrate the diverse chemistries that have been employed to produce supramolecular healable gels in this young, but growing field of research. Clearly the low physical strength of gels restricts their use to very specialized applications. However, efforts have been made to increase the strength of these materials by producing composite structures that have greatly increased the mechanical properties and are covered in a subsequent section (see Section 4.9.4).

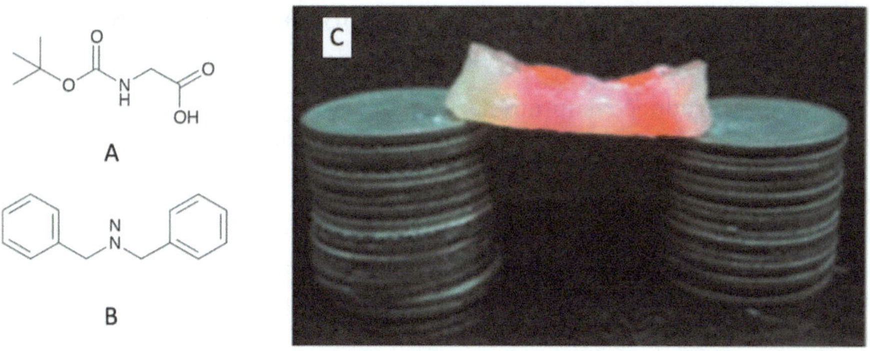

Figure 4.31 Structures of Boc-protected glycine (A) and dibenzylamine (B) used to form freestanding organogels. (C) Example of a self-supporting sample of a healed gel produced from joining 5 pieces of gel (2 sections dyed pink for clarity). Reproduced from ref. 58 with permission from John Wiley & Sons, Copyright 2012 Wiley-VCH Verlag GmbH & Co. KGaA, Weinheim.

4.9 Healable Supramolecular Polymeric Nanocomposite Materials

It is clear from the broad range of supramolecular motifs and polymeric structures employed within supramolecular healable materials in this chapter that significant scientific progress has been made in this field over the past 20 years. Many of these materials are relatively weak, and frequently have T_g values well below room temperature. Whilst these properties are not an insurmountable drawback for many applications, they are not suitable to replace the high strength polymers now being employed in safety critical situations, for example, in the airframes of aircraft or body panels of cars. However, the ability to repair these load bearing components would clearly be advantageous in reducing maintenance costs and extending the lifetime of such products.

To increase the strength of these supramolecular materials, chemists have investigated the ability of nano-composite materials to undergo healing.[60] Composite materials are composed of at least two distinct materials, and the combination of these produce a final product with enhanced physical characteristics when compared to either of the separate components. Carbon or glass fiber reinforced epoxy composites have become widely used as high-strength, light-weight materials over the past 50 years.[61] However, the covalently cross-linked nature of the matrix polymer when combined with the macroscopic scale of such fiber fillers does not permit 'healing' through the mechanisms described in this chapter. In contrast, if the filler component is nano-scale in its dimensions and embedded within a supramolecular polymeric continuous phase, then healing can be achieved with modest energy input as well as on useful timescales.

4.9.1 Healable Supramolecular Nanocomposite Polymers Containing π–π Stacking

Fox *et al.*[62] produced supramolecular healable nano-composite materials using the healable materials containing π–π stacking integrations between NDI chain folding components and pyrenyl derivatives discussed in Section 4.4. In this study the properties of extremely ridged cellulose nanocrystals (CNCs) were exploited which have a tensile modulus of up to 140 GPa.[63] A series of formulations were produced with loading levels up to 20 wt% CNCs. The tensile modulus of the resulting nano-composites was found to increase from *ca.* 8 MPa for the pristine polymer blend without the filler to *ca.* 261 MPa for the sample that contained 10% of the CNC filler (Figure 4.32(A)). This enhancement in strength was in good agreement with predictions using a classic percolation model for dispersal, suggesting that the CNCs are homogeneously distributed within the polymer.[63,64] However, above 10 wt% CNC loading the resulting films were non-homogeneous with phase separation occurring between the continuous phase and the filler, resulting in a

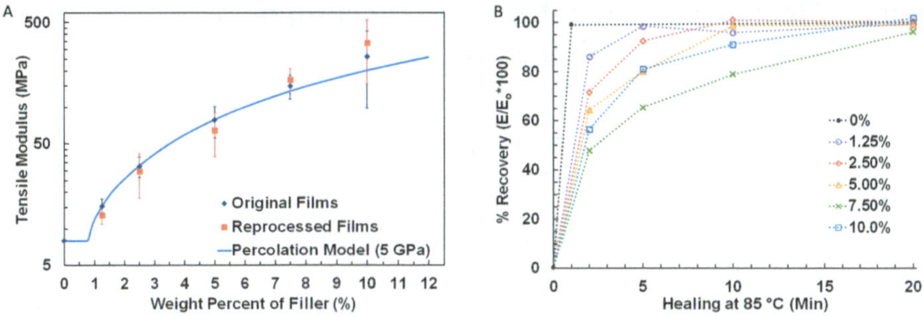

Figure 4.32 Plot of (A) variation in the strength of the pristine and healed supramolecular nanocomposite materials as a function of wt% CNC filler. The blue line is the predicted strength of the materials as calculated by the percolation model. (B) Healing efficiency as a function of healing time for nanocomposites with varying loading levels of CNCs. Reproduced from ref. 62 with permission from American Chemical Society, Copyright 2012.

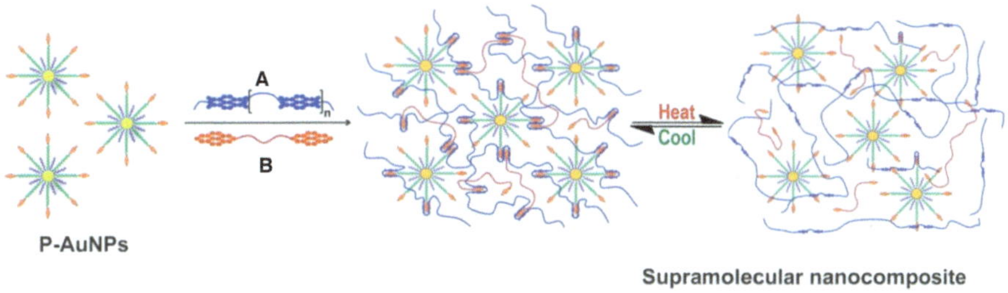

Figure 4.33 Schematic of the incorporation of pyrene functionalized AuNPs (P-AuNPs) into a healable polymer blend containing polymers with NDI (A) and pyrene (B). Adapted from ref. 65B with permission from the Royal Society of Chemistry.

drop in mechanical performance. The healing efficiencies for each of the samples with less than 10 wt% CNC were approximately 100% after 20 minutes at 85 °C, although there was variation in the rate of healing between the samples (Figure 4.32(B)).

Using related polymeric systems, Hayes and co-workers functionalized the surface of gold nano-particles (AuNPs) with π-electron rich pyrenyl residues (P-AuNPs, Figure 4.33).[65] The P-AuNPs were then used as the filler component within an NDI-based polymer blend nanocomposite. In this system the P-AuNPs were additional cross-linking nodes in the system.[65b] Samples with 10% AuNPs exhibited significantly higher tensile moduli than the base polymer alone (0.2 MPa *vs.* 0.6 MPa) whilst both materials exhibited essentially 100% healing efficiency. Reference nano-composites with AuNP fillers that lacked the pyrenyl surface groups were also produced. They exhibited a significant (*ca.* 50%) reduction in film strength, clearly demonstrating the importance of the designed non-covalent interactions on the physical properties of the system.

4.9.2 Healable Supramolecular Nanocomposite Polymers Containing Hydrogen Bonding Interactions

Foster, Weder and co-workers produced healable supramolecular nanocomposites where the continuous phases included a polymer that contained hydrogen interactions.[66] In this study, CNCs were functionalized with UPy groups to enable good dispersion within a supramolecular polymer matrix (Figure 4.34). Films containing up to 20 wt% of the modified CNC filler were found to be non-phase separated and exhibited tensile moduli approximately an order of magnitude greater than the bulk polymer (13 MPa *vs.* 158 MPa). These nanocomposites could be healed by exposure to light after only 80 seconds (wavelength: 320–390 nm, power: 250 mW cm^{-1}) with essentially 100% healing efficiency.

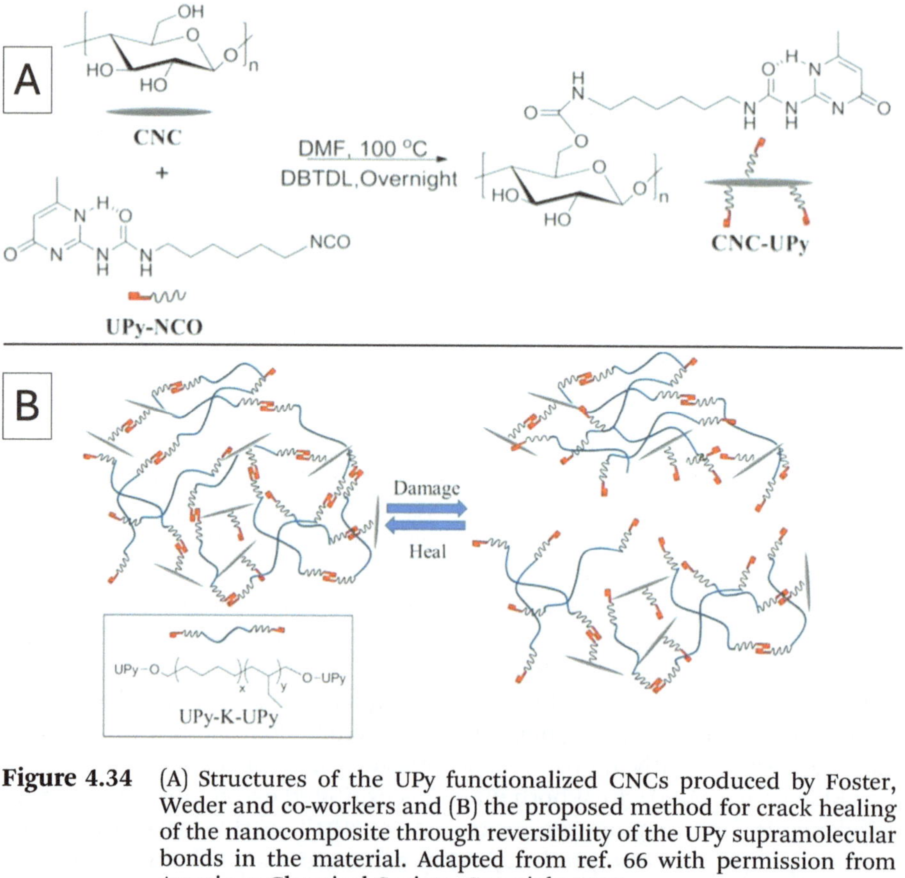

Figure 4.34 (A) Structures of the UPy functionalized CNCs produced by Foster, Weder and co-workers and (B) the proposed method for crack healing of the nanocomposite through reversibility of the UPy supramolecular bonds in the material. Adapted from ref. 66 with permission from American Chemical Society, Copyright 2013.

4.9.3 Healable Supramolecular Nanocomposite Polymers Containing Metal–Ligand Interactions

Gong and co-workers investigated a healable metallosupramolecular nanocomposite polymer system containing functionalized carbon nanotubes (CNs) as the fillers.[67] In this study, commercially available hydroxy functionalized CNs (CN–OH) were modified with a polyurethane that was terminated with terpyridine ligands, which are known to form stable complexes with a range of metal ions. The total loading of CN–OH was only 0.5% of the weight of the nanocomposite (Figure 4.35). Stress–strain tests on the nanocomposites revealed that the material was elastomeric in nature and could be extended to 658% strain before breaking. On addition of Zn^{2+} ions to link the CNs together, the maximum strain exhibited by the sample increased to 1076%, whilst the modulus of toughness increased from 51 to 141 MPa. Healing tests were carried out by scratching the surfaces (4 mm in length *ca.* 50 μm depth) and summitting the cut films to various healing stimuli prior to strength testing. In comparison to cut but unhealed samples, the nanocomposite was able to regain its strength with remarkable healing efficiencies (93% or greater) in response to exposure to near IR light, heat or solvent.

4.9.4 Healable Nanocomposite Gels

Nanoclays have been extensively studied as fillers in polymer composites for over 25 years.[68] Although there are many types of clay (*i.e.* montmorillonite,

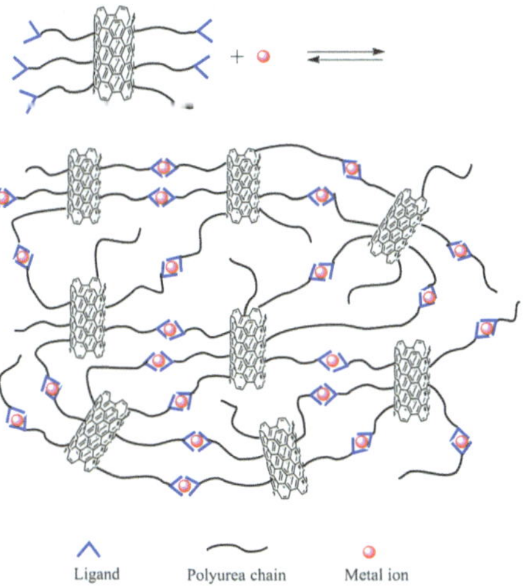

Figure 4.35 Schematic of the metallosupramolecular CN nanocomposite produced by Gong and co-workers. Adapted from ref. 67 with permission form the Royal Society of Chemistry.

hectorite, laponite, *etc.*) they are all principally composed of aluminosilicate layers, which must be separated (exfoliated) into platelets that are just a few nanometers thick in order to be homogeneously dispersed throughout the polymer matrix. In a series of papers beginning in 2002, Haraguchi and co-workers demonstrated that it was possible to harness the same nanoclay fillers to increase the strength of conventional covalent hydrogel polymers (for example, poly(*N*-isopropylacrylamide).[69]

The first example of a healable nano-composite gel was reported by Aida and co-workers in 2010.[70] The group produced a multi-component hydrogel consisting of clay nano-sheets, sodium polyacrylate (Figure 4.36(A)) and guanidium functionalized dendrimers of varying generations (Figure 4.36(B)).

The organic components were designed for separate tasks, and the sodium polyacrylate performed a key role in exfoliating the nanoclay through favorable interactions with the polar groups on the surface of the clay platelets (Figure 4.37(b)). The resulting polymer-coated nanoclay was then cross-linked through transient salt bridges formed between the positively charged guanidium residues on the dendrimers and the acid groups on the linear polyacrylate (Figure 4.37(c)), generating a three-dimensional supramolecular network. The strength of the hydrogels was found to increase with increasing number of surface groups on the dendrimers (by increasing the generation number from 1 to 3) with the strongest gels exhibiting a storage modulus of approximately 0.5 MPa (see Section 1.5.3).

The healing ability of the nano-composite hydrogel was demonstrated by rheological studies (see Section 1.5) and visually by placing cubes of gel in contact to form self-supporting structures (see Figure 4.38). In common with the supramolecular rubber reported by Leibler and co-workers[28] (Section 4.3) separation of cut sections (for more than a minute) resulted in loss of healable characteristics as a consequence of the rapid re-arrangement of the

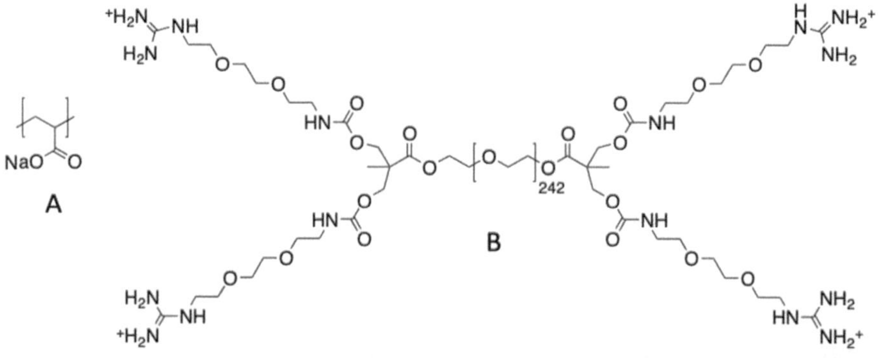

Figure 4.36 The structure of sodium polyacrylate (A) and the first generation of guanidium functionalized dendrimer (B) used to produce healable nanoclay composite hydrogels.

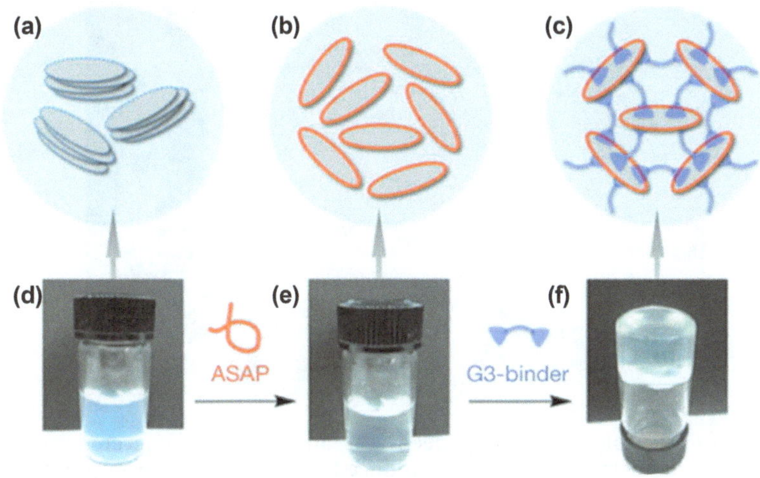

Figure 4.37 Schematic representation of undispersed nanoclay platelets, exfoliated sodium polyacrylate (ASAP) coated nanoclay platelets and the dendrimer cross-linked supramolecular gel (a to c, respectively). Photographs of aqueous samples of undispersed nanoclay platelets, exfoliated sodium polyacrylate (ASAP) coated nanoclay platelets and the dendrimer crosslinked supramolecular gel (d to f, respectively). Adapted from ref. 70 with permission from Springer Nature, Copyright 2010.

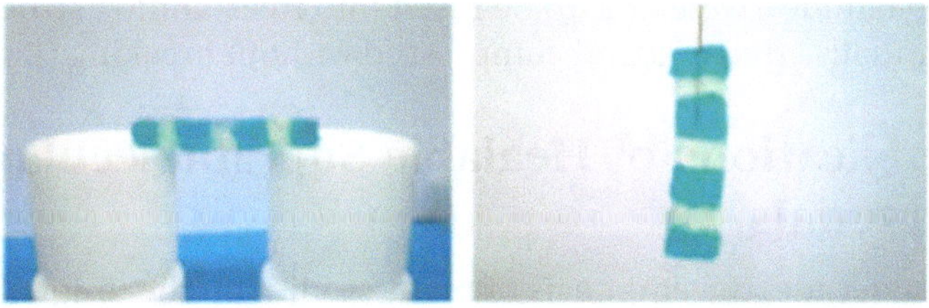

Figure 4.38 Photographs of a 'healed' sample of free standing nanoclay hydrogel formed by fusing 7 sections of freshly cut gel (4 sections dyed blue for clarity). Reproduced from ref. 70 with permission from Springer Nature, Copyright 2010.

supramolecular motifs from the cleaved surface into the bulk over short time periods.

Haraguchi and co-workers also deployed nanoclays as the filler component to construct two healable composite hydrogels which differed according to the structure of the polymeric components.[71] These systems were produced by free radical polymerization of either *N*,*N*-dimethylacrylamide or *N*-isopropylacrylamide in water containing varying quantities of the nanoclay to generate free-standing gels with approximately 10–15 wt% total solid content. Visually impressive healing of deep cuts in the gels was observed by leaving the samples at 37 °C for 48 hours (Figure 4.39). Further studies

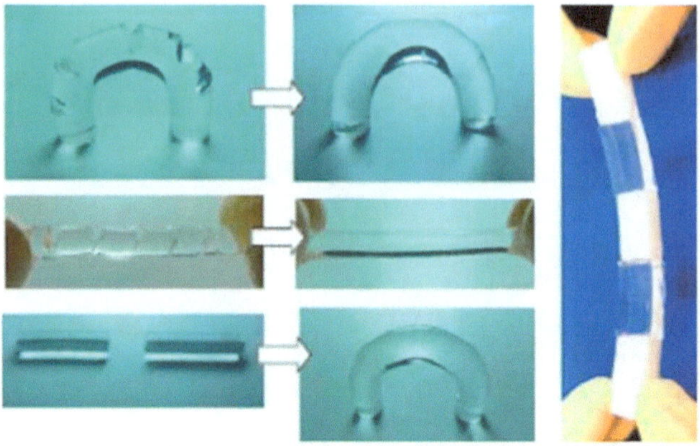

Figure 4.39 Examples of healing deep cuts and completely severed samples of nanoclay hydrogels. Reproduced from ref. 71 with permission from John Wiley & Sons, Copyright 2011 WILEY-VCH Verlag GmbH & Co. KGaA, Weinheim.

showed that healing efficiency was related to the contact time of the severed edges, and η_{eff} increased from 27 to 57 to almost 100% at healing times of 1, 4 and 10 hours, respectively (at 50 °C in a sealed container). Unusually for supramolecular systems, healing was observed even after the severed surfaces had been separated for several days. In addition, sections of hydrogen composed of different polymers (poly(*N*,*N*-dimethylacrylamide) or poly(*N*-isopropylacrylamide)) were able to be fused into freestanding structures that could be manually stretched and compressed without breaking (Figure 4.39).

4.10 Applications of Healable Supramolecular Materials

Over the past decade, the emphasis of many studies on the unique attributes of healable supramolecular polymers has shifted towards their incorporation into multiple real-world applications. Frequently, each application uses the responsive nature of the supramolecular bond to confer additional functionality over that possible using conventional high molecular weight polymers or covalently cross-linked systems. Such is the pace of the field that a comprehensive survey of all these new discoveries is beyond the scope of this chapter, but the exemplars selected in the following sections are chosen to provide a flavor of the impact that healable supramolecular polymers may have on society in the upcoming years.

4.10.1 Debondable and Rebondable Adhesives

The manufacture of complex products ranging from mobile phones to automobiles requires that materials with disparate mechanical and surfaces properties must be joined together to make a single functioning unit.

Traditional fixing modes such as wields, rivets and screws are not suitable to join together every permutation of material, and increasing frequently, bonding using adhesives[72] is a more rapid and scalable method for achieving this goal. Moreover, during the past 10 years an ever-greater drive to account for the whole lifecycle of the product in the original plan of the manufacturing process has become important. Often driven by consumer demand, manufacturers are seeking to futureproof their products by extending their lifecycle or facilitating recycling. These goals can only be achieved by building the product so that the individual parts can be rapidly disassembled on demand to be replaced, reused or recycled in an appropriate waste stream. Producing adhesives which harness the reversible nature of the supramolecular bond has been one of the key developments in the field of debondable and rebondable adhesives.[73]

4.10.1.1 *Stimuli for Debonding/Rebonding Adhesives*

As described in the previous sections of this chapter, the range of stimuli that can impact on the mechanical properties of supramolecular polymers is varied. Therefore, a fundamental factor in the design of responsive adhesives is careful selection of the stimulus to ensure it is compatible with the end use. A commercial hot melt adhesive can be induced to debond by the application of thermal energy,[72] but this will not be appropriate if the joint to be debonded is concealed within the product, or if the bonded substrates are not stable at the required debonding temperature. Thermal instability could be overcome by using a chemo- or photo-responsive supramolecular polymer, but again, this must be tensioned with the necessity for the appropriate chemical stimulus to reach the adhesive or the bonded substrates being transparent at the required wavelength to enable debonding. Thus, a range of responsive adhesives have been developed which are described in the following sections.

4.10.1.2 *Thermally Healable Adhesives*

The inverse relationship between temperature and the association constant of a supramolecular bond is one of the fundamental attributes of these systems. Indeed, commercial hot melt adhesives are frequently polyurethane (PU) based solids at room (or bonded) temperature but are applied at an elevated temperature where the polymer flows sufficiently to allow efficient coating of the substrate surfaces.[72] During the heating/cooling process no new covalent bonds are formed and the strength of the solid (cooled) state is highly dependent on the hydrogen bonding interactions between the PU chains.

Thermal debonding of supramolecular adhesives which contained the well-studied UPy group formed some of the early ground work in this field. Long and co-workers synthesized polybutylacrylate (PBA) systems featuring up to 10.4 mol% UPy groups as pendant receptor units (Figure 4.40).[74] Peal

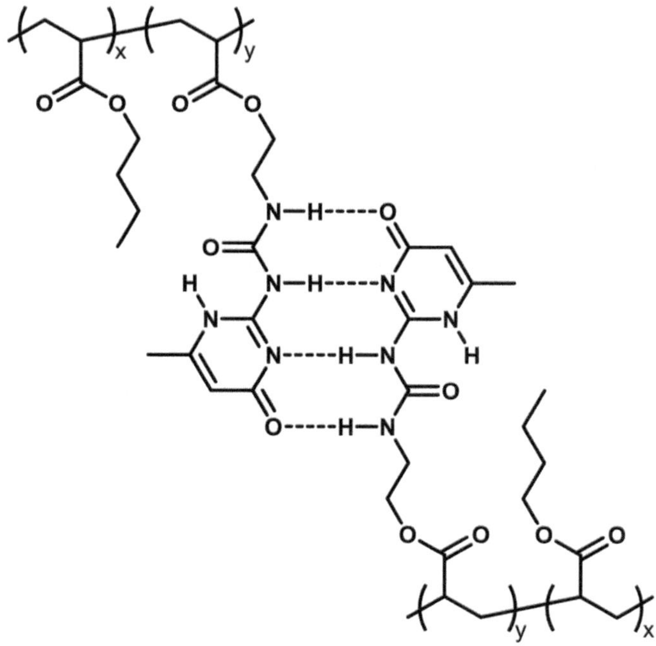

Figure 4.40 Structures of the UPy functionalized acrylate polymers produced by Long and co-workers. The supramolecular hydrogen bonding interactions are shown as dashed lines. Reproduced from ref. 75, https://doi.org/10.1016/j.reactfunctpolym.2022.105209, under the terms of the CC BY 4.0 license https://creativecommons.org/licenses/by/4.0/.

strength tests showed that whilst PBA and the UPy functionalized PBAs bonded to glass substrates, the addition of the pendant UPy groups increased the peal strength three-fold. This result showed that the addition of the UPy receptor groups aided adhesion to the glass substrate.

Subsequent studies by Bosman and co-workers focused on the adhesive nature of a UPy functionalized PDMS (Figure 4.41).[76] This polymer adhered to a glass substrate whilst supporting a static load of 1 kg without breaking. However, if deliberately fractured the joint could be reformed by heating the adhesive covered surfaces to 120 °C and rebonding them. The healed joint was found to have essentially the same strength as the pristine bond.

Greenland and co-workers studied a series of PUs that could undergo thermoreversible-bonding and debonding.[77] Moreover, the polymer contained a chemo-responsive functionality that resulted in covalent bond breakage (Figure 4.42(A)). This design allows for both reversible bonding by dissociation of the supramolecular hydrogen bonds between polymer chains and one-time non-reversible debonding by permanent reduction in the molecular weight of the polymer backbone caused by exposure to fluoride ions. During initial studies, a PU based polymer (Figure 4.42(B)) was synthesized that was able to bond a range of substrates including PVC, wood,

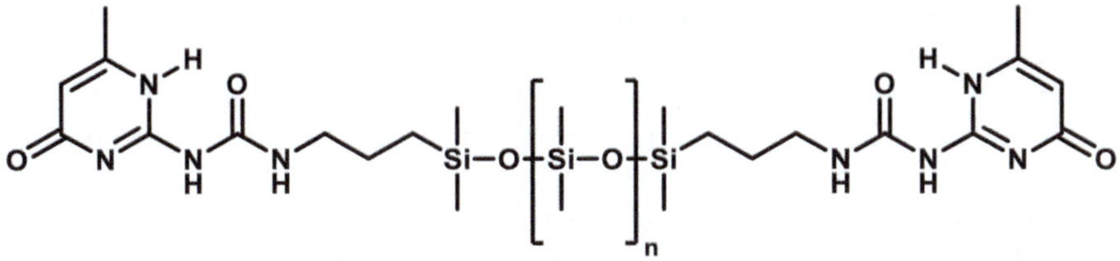

Figure 4.41 Structure of the UPy functionalized PDMS adhesive produced by Bosman and co-workers that could rebond fractured joints at 120 °C. Reproduced from ref. 75, https://doi.org/10.1016/j.reactfunctpolym.2022.105209, under the terms of the CC BY 4.0 license https://creativecommons.org/licenses/by/4.0/.

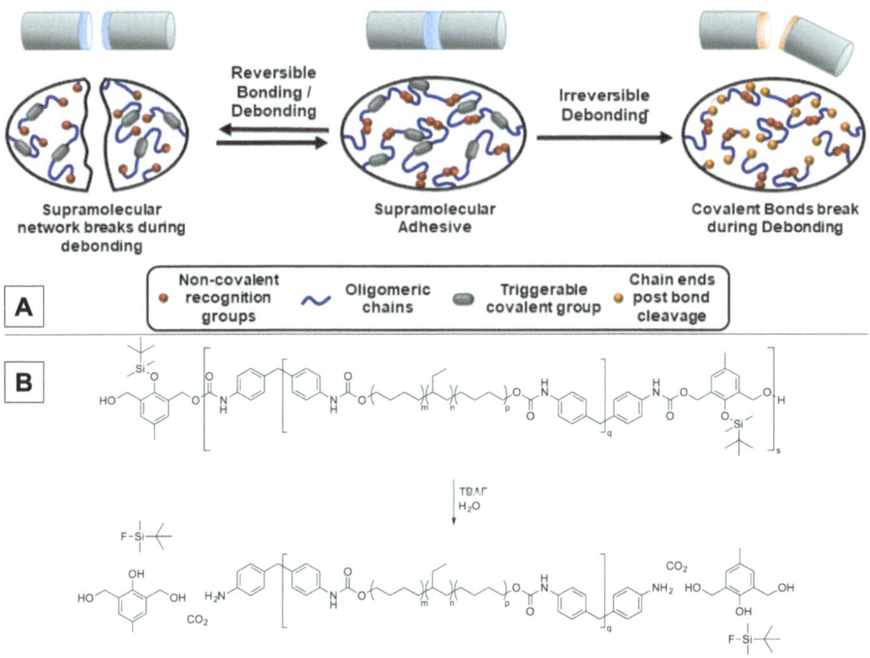

Figure 4.42 (A) Schematic of the thermoreversible adhesion combined with fluoride responsive depolymerization of the PU based adhesive produced by Greenland and co-workers,[75] and (B) proposed fluoride induced degradation pathway for the adhesive which results in a permanent reduction of adhesion. Adapted from ref. 77*a* with permission from the Royal Society of Chemistry.

aluminum and glass. The adhesive was able to undergo up to 5 break/heal cycles without loss of bonding strength after rebonding at 160 °C. Immersion in an acetone solution of tetrabutyl ammonium fluoride (TBAF, a source of fluoride ions) resulted in the desired irreversible breaking of the polymer backbone facilitated by removal of a silyl ether group (Figure 4.42(B)). The consequent reduction in molecular weight from *ca.* 26 kg mol^{-1} to 6.2 kg mol^{-1} gave a reduction in adhesion strength of approximately 40%.

Later work by the same group showed that the bonding temperature could be tuned by manipulating the chemical composition of the polymer backbone.[78b] Polymers containing polyesters in the PU backbone were able to bond and rebond at 49 °C, whilst retaining bonding strength, a significantly lower temperature than that of the analogous materials containing hydrogenated polybutadiene (160 °C). By varying the structure of the silylether component that enabled depolymerization, it was also possible to produce cross-linked rather than linear variants of this multifunctional adhesive.[78]

4.10.1.3 *Thermally Healable Adhesives Not Using Conventional Heating Methods*

Weder and co-workers reported work on two distinct methods for producing adhesives that were able to bond and rebond using UV radiation or heat as the stimulus. In 2014, a telechelic UPy functionalized polybutadiene was reported that was blended with the dye Tinuvin 326.[79] Tinuvin 326 absorbs UV radiation and converts it into heat, raising the temperature of the adhesive and dissociation of the UPy dimers (Figure 4.43(A)). The material showed reversible adhesive properties when bonding glass substrates without loss of strength both *via* light-induced heating and direct thermal stimuli. In the same paper, a metallosupramolecular polymer was disclosed that contained benzimidazole pyridine ligands and Zn(II) ions (Figure 4.43(B)) also behaving as an adhesive.[79] This material debonds from its substrates upon exposure to UV light, yet on removal of the radiation source, the metal–ligand interactions reform, and the original strength of the bond returns.

As discussed in Section 4.9, the addition of nano-scale fillers to supramolecular polymers can dramatically improve their strength. It is also possible to add fillers with a dual function, to further enhance the properties of the material. For example, addition of Fe_3O_4 nanoparticles not only improves the tensile modulus and modulus of toughness of polymers, but the particles can undergo hysteresis heating whereby they can convert energy from an oscillating magnetic field into heat to promote bonding/debonding (Figure 4.44). Thermoresponsive nanocomposite adhesives have been produced that use this methodology to deliver quicker bonding times in comparison to traditional thermal heating of the adhesive.[80] However, reversing this method to induce debonding has only recently been reported.[81] Greenland and co-workers showed that a PU/Fe_3O_4 nanocomposite adhesive exhibited excellent bonding of metals, woods and plastics. Moreover, the nanocomposite adhesive underwent hysteresis heating resulting in debonding of glass substrates after less than 1 minute of being subject to an alternating magnetic field.[81a]

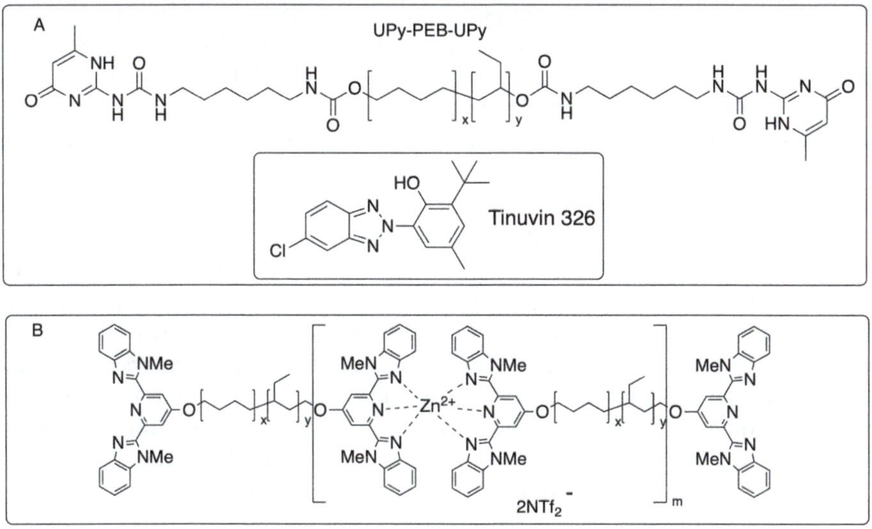

Figure 4.43 (A) Structure of the telechelic UPy functionalized supramolecular adhesive containing the dye Tinuvin 326 produced by Weder and co-workers.[79] (B) A metallosupramolecular polymeric adhesive that can debond in response to heat or UV radiation.[79]

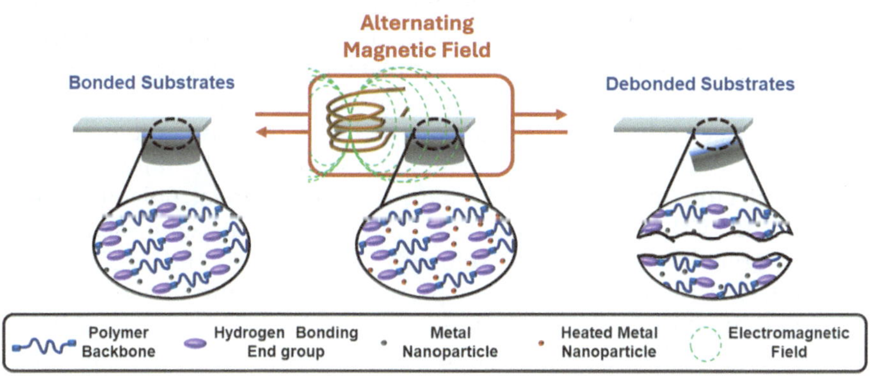

Figure 4.44 Schematic of Fe_3O_4 nanocomposite adhesives that de/rebond on demand when placed in an alternating magnetic field. Adapted from ref. 75.

4.10.2 Self-healing Supramolecular Polymers for Battery Technology

One of the key technologies underpinning life in the 21st century has been the widespread instruction of high energy density electrical storage solutions.[82] Within this varied field, lithium ion batteries (LIBs) have come to dominate the market for personal electronics (laptops, smart phones) or transportation (electric cars). An in-depth knowledge of the structure of these batteries is beyond the scope of this report, however, electrical storage devices are generally a sandwich type structure whereby the cathode and anode are separated by an electrolyte packing material. Each of these components can contain polymeric materials which can have a multitude of functions from supporting the charge carriers to binding components or sealing the power cell.

Most people will have experienced the phenomena of the storage capacity of LIBs decreasing as the number of charge/discharge cycles increases. The physical reason for the reduction in storage capacity of LIBs is that the battery undergoes a significant volume change during the charge and discharge cycle, which causes cracks and damage inside the cell, which can ultimately lead to short circuiting and even fire.[82] Therefore, over the past 10 years there have been significant efforts centered on producing polymeric materials that are capable of healing damage within the cell whilst also withstanding the demanding chemical environment found within in LIBs. Achieving these goals would significantly extend the useful life of these batteries.

4.10.2.1 Supramolecular Self-healing Polymers as Binders in the Anodes of Batteries

In 2013, Boa and co-workers presented a solution to the problem associated with degradation of the brittle silicon-based anodic electrode components in batteries.[83] Conventional rechargeable battery design requires that particles of silicon are bound together with a polymer to the current collector to maintain the electrical contact during the charge/discharge cycle. During repeated charge-cycling, changes in volume cause the inorganic particles to crack and disperse, reducing the contact efficiency of the anode with the current collector. In their novel system, the conventional polymer bonder was replaced with a supramolecular self-healing material based on the hydrogen bonding system developed by Leibler and co-workers[28] (Section 4.3). Although this did not prevent the cracking of the brittle silicon electrode, it did help maintain good contact of the electrode with the current collector. Thus, cells containing the electrode with the self-healing polymer maintained at least 80% of their maximum charge after 90 cycles when compared to just 10 cycles for the system with the traditional 'non' healing binder (Figure 4.45).

4.10.2.2 Supramolecular Self-healing Polymers in Battery Electrolytes

Other groups have been working on the incorporation of self-healing functionalities into the electrolyte components of batteries. Huang and co-workers produced sodium polyacrylic acid (PANa)-based hydrogels for use in solid-state aqueous rechargeable NiCo||Zn batteries.[84] Typically, PANa hydrogels are hard to heal if broken. However, the addition of Fe^{3+} ions to the gel results in

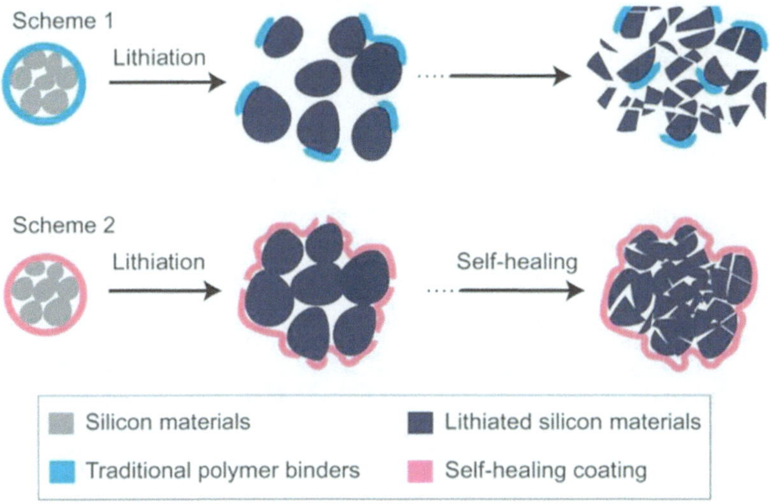

Figure 4.45 Schematics of the expansion and degradation of the silicon electrons with transitional binders and self-healing supramolecular binders. Adapted from ref. 83 with permission from Springer Nature, Copyright 2013.

the formation of supramolecular cross-links facilitated by Fe^{3+} ions complexing with the carboxylate groups on different polymer chains (Figure 4.46(A)). Thus, when cut, these cross-links are able to reform across the fracture void without significant loss of charge storing ability. Indeed, batteries created with this electrolyte were able to undergo 4 break heal cycles whilst retaining over 80% of their initial performance (Figure 4.46(B)).

4.10.3 Wearable Technologies Employing Supramolecular Healable Polymers

There are a range of current and emerging technologies that rely on materials that can either fix to, or be worn on the skin whilst the user continues with their everyday activities. In many of these devices the polymeric component should flex with the motion of the body and are likely to become damaged or abraded during use. In these situations, the low T_g and relatively weak nature of many supramolecule materials is not disadvantageous. The generation of healable versions of these supramolecular-based technologies would be useful to extend the working lifetime of the product, as detailed with the selected exemplars described below.

4.10.3.1 Healable Supramolecular Polymers for Wound Dressings

In 2016 Hayes and co-workers produced a supramolecular PU that adhered to the skin and was able to heal fractures at body temperature.[85] The

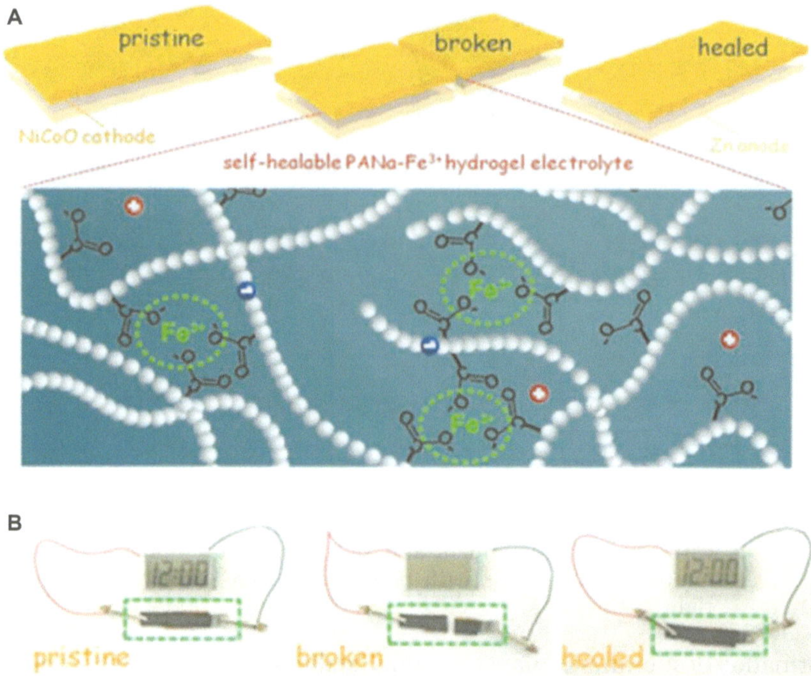

Figure 4.46 (A) Schematic of the break heal process of the electrolyte with the expansion showing the addition of the Fe^{3+} ions cross-linking PAN chains. (B) An experiment showing how the conductivity of the electrolyte could be regained after healing even after complete fracture of the sample. Adapted from ref. 84 with permission from John Wiley & Sons, Copyright 2018 Wiley-VCH Verlag GmbH & Co. KGaA, Weinheim.

polymer was based[86] on a hydrophobic diol that was functionalized with methylene diphenyldiisocyanate and subsequently end-capped with a hydrogen bonding end group (4-(2-aminoethyl)morpholine) (Figure 4.47). This polymer was shown to be non-toxic and adhered to pig skin (a readily available human skin model) and it was proposed that it could be used as a wound dressing. Healing studies carried out at 37 °C showed that the flat surface profile of the material fractures could be restored after fracture within 2 hours even when the polymer remained on the skin.

4.10.3.2 Supramolecular Healable Polymers in Wearable Electronics

Sensing changes in the body in real time offers the potential for more accurate diagnostics as well as seamless integration with, and control of, computers and robots.[87] In 2019, Zeng and co-workers created a conducting

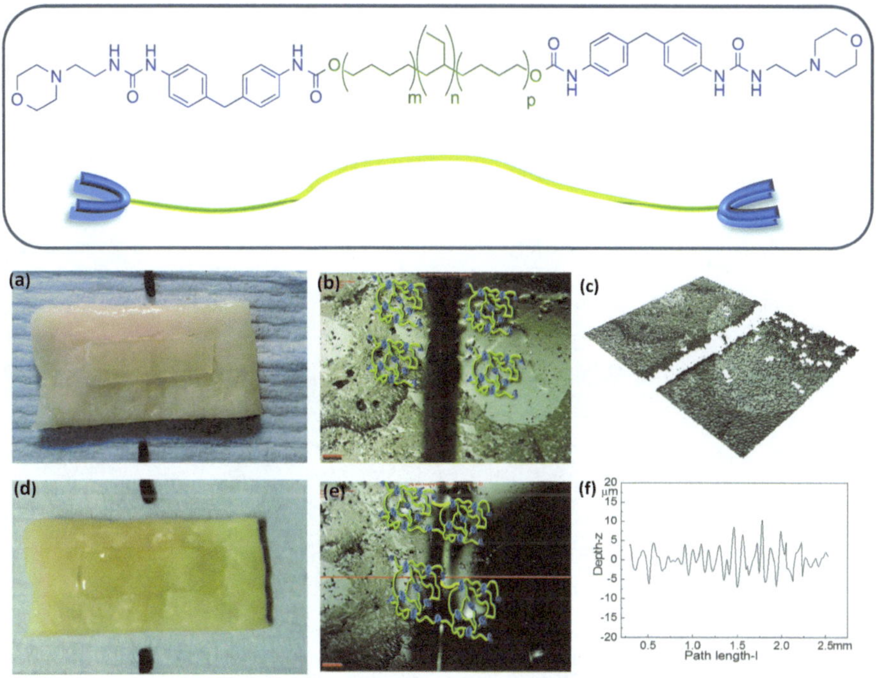

Figure 4.47 (Top) Chemical structure and schematic of the healable supramolecular wound dressing produced by Hayes and co-workers. (Bottom) (a) Sample of the fractured polymer on pig skin; (b) microscope image of the fracture site; (c) 3D profilometry of the fracture site; (d) photograph of the skin after healing; (e) microscope image of the skin after healing; (f) surface topography following the red line in (e) across the healed fracture site showing the removal of the depression at the break. Adapted from ref. 85 with permission from the Royal Society of Chemistry.

material based on a UPy functionalized acrylate polymer (Figure 4.48(A)).[88] This was in turn fabricated into a conducting composite material by dissolving the pre-formed polymer in ethanol containing pyrrole, which was then polymerized by the addition of ammonium persulfate (APS) and a copper salt. Finally, samples for testing were formed by solution casting. Composite materials containing polypyrrole and just 5 mol% UPy residues could be extended to 10 times their original length without breaking. The polymeric films were conductive and could undergo break–heal tests by manually pressing freshly cut edges together and leaving them for five minutes at room temperature, a process that did not inhibit the conductivity of the material (Figure 4.48(B)). The resistivity of the film varied monotonically with the applied strain between 0 to 300%. Therefore, the material could behave as a strain sensor. This attribute was demonstrated by attaching a sample to the skin on the forearm where it was able to sense the pulse in the radial artery (Figure 4.48(C)).

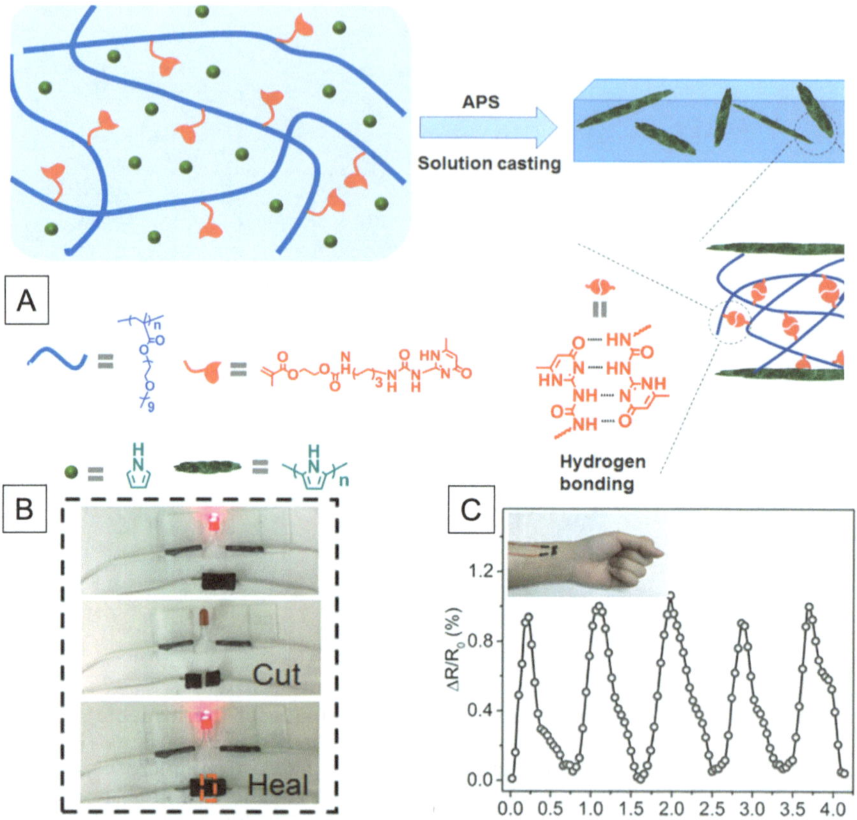

Figure 4.48 (A) Chemical structures and schematics of the supramolecular healable wearable sensor produced by Zeng and co-workers; (B) cut/heal tests of the material showing regeneration of conductivity and (C) resistivity measurements of the polymer on the wrist showing the pulse sensing. Adapted from ref. 88 with permission from American Chemical Society, Copyright 2019.

4.11 Conclusions and Future Perspectives

This chapter has used examples from the literature to demonstrate the key design criteria that must be fulfilled in order to produce healable supramolecular polymeric systems. Although the strength and directionality of the supramolecular interactions harnessed within each of these materials is of great importance, in the solid state, crystallinity and phase separation of the components can also be engineered to build in additional performance and functionality.

It is proposed that in supramolecular structures, fractures occur through scission of the weaker supramolecular interactions rather than the stronger covalent bonds. Healing is facilitated by re-engagement of the supramolecular motifs across the fracture void. However, these universal design elements may also impose an upper limit on strength and, therefore, scope of application of some of these materials, for example:

i. Materials using supramolecular motifs may always fail before conventional crosslinked materials as a consequence of their lower bond energies.
ii. In order to allow enough molecular motion for healing to occur on a reasonable timescale the phase of the material containing the supramolecular motif must be at a greater temperature than its T_g (or T_m), or in the gel phase.

These requirements result in the high proportion of weak, elastomeric or gel type materials detailed throughout this chapter. Increasing the strength of the supramolecular interactions has been demonstrated to result in enhanced mechanical properties, but this comes with the requirement for more forceful healing conditions. Whilst healable nanocomposites may offer a path to greatly enhanced strength, their use has generally been confined to lab-based systems, far from real world applications. Over the most recent years attention has turned towards identifying the first commercial uses of these novel systems. This was exemplified by the implementation of supramolecular healable materials in a range of potentially desirable environments including being used in adhesives, as binders in battery electrolytes and incorporated into wearable electronic technologies.

References

1. For a historic perspective see: S. C. Rasmussen, *Angew. Chem., Int. Ed.*, 2021, **60**, 8012.
2. S. D. Bergman and F. Wudl, *J. Mater. Chem.*, 2008, **18**, 41.
3. For reviews see: (a) S. Burattini, B. W. Greenland, D. Chappell, H. M. Colquhoun and W. Hayes, *Chem. Soc. Rev.*, 2010, **39**, 1973; (b) E. B. Murphy and F. Wudl, *Prog. Polym. Sci.*, 2010, **35**, 223; (c) M. W. Urban, *Prog. Polym. Sci.*, 2009, **34**, 679; (d) B. J. Blaiszik, S. L. B. Kramer, S. C. Olugebefola, J. S. Moore, N. R. Sottos and S. R. White, *Ann. Rev. Mater. Res.*, 2010, **40**, 179; (e) R. P. Wool, *Soft Matter*, 2008, **4**, 400; (f) S. Seiffert and J. Sprakel, *Chem. Soc. Rev.*, 2012, **41**, 909; (g) D. Y. Wu, S. Meure and D. Solomon, *Prog. Polym. Sci.*, 2008, **33**, 479; (h) J. A. Syrett, C. R. Becer and D. M. Haddleton, *Polym. Chem.*, 2010, **1**, 978; (i) S. Y. Wang and M. W. Urban, *Nature Rev. Mater.*, 2020, **5**, 583; (j) D. Y. Xia, P. Wang, X. F. Ji, N. M. Khashab, J. L. Sessler and F. H. Huang, *Chem. Rev.*, 2020, **120**, 6070.
4. K. Jud, H. H. Kausch and J. G. Williams, *J. Mater. Sci.*, 1981, **16**, 204.
5. Y. H. Kim and R. P. Wool, *Macromolecules*, 1983, **16**, 1115.
6. A.-D. N. Celestine, N. R. Sottos and S. R. White, *Polymer*, 2015, **69**, 241.
7. A. P. Esser-Kahn, S. A. Odom, N. R. Sottos, S. R. White and J. S. Moore, *Macromolecules*, 2011, **44**, 5539.

8. S. R. White, N. R. Sottos, P. H. Geubelle, J. S. Moore, M. R. Kessler, S. R. Sriram, E. N. Brown and S. Viswanathan, *Nature*, 2001, **409**, 794.
9. (a) K. S. Toohey, N. R. Sottos, J. A. Lewis, J. S. Moore and S. R. White, *Nat. Mater.*, 2007, **6**, 581; (b) A. P. Esser-Kahn, P. R. Thakre, H. F. Dong, J. F. Patrick, V. K. Vlasko-Vlasov, N. R. Sottos, J. S. Moore and S. R. White, *Adv. Mater.*, 2011, **23**, 3654.
10. L. Yang, X. Tan, Z. Wang and X. Zhang, *Chem. Rev.*, 2015, **115**, 7196.
11. A. Ciferri, *Supramolecular Polymers*, Marcel Dekker, New York, 2000.
12. L. Brunsvel, B. J. B. Folmer, E. W. Meijer and R. P. Sijbesma, *Chem. Rev.*, 2001, **101**, 4071.
13. R. P. Sijbesma, F. H. Beijer, L. Brunsveld, B. J. B. Folmer, J. H. K. K. Hirschberg, R. F. M. Langer, J. K. L. Lowe and E. W. Meijer, *Science*, 1997, **278**, 1601.
14. T. Park, S. C. Zimmerman and S. Nakashima, *J. Am. Chem. Soc.*, 2005, **127**, 6520.
15. C. Fouquey, J.-M. Lehn and A.-M. Levelut, *Adv. Mater.*, 1990, **2**, 254.
16. C. R. Martinex and B. L. Iverson, *Chem. Sci*, 2012, **3**, 2191.
17. J.-M. Lehn, *Angew. Chem., Int. Ed.*, 1988, **100**, 91.
18. C. J. Pederson, *Angew. Chem., Int. Ed.*, 1988, **100**, 1009.
19. D. J. Cram, *Angew. Chem., Int. Ed.*, 1988, **100**, 1021.
20. M. Barboiu, G. Vaughan, R. Graff and J.-M. Lehn, *J. Am. Chem. Soc.*, 2003, **125**, 10257.
21. For a review of supramolecular polymerizations see: J. D. Fox and S. J. Rowan, *Macromolecules*, 2009, **42**, 6823.
22. A. W. Bosman, R. P. Sijbesma and E. W. Meijer, *Mater. Today*, 2004, **7**, 34.
23. (a) L. L. de Lucca Freitas and R. Stadler, *Macromolecules*, 1987, **20**, 2478; (b) C. Hilger and R. Stadler, *Macromolecules*, 1990, **23**, 2097; (c) C. Hilger, R. Stadler and L. L. de Lucca Freitas, *Polymer*, 1990, **31**, 818; (d) C. Hilger and R. Stadler, *Polymer*, 1991, **32**, 3244; (e) C. Hilger, M. Dräger and R. Stadler, *Macromolecules*, 1992, **25**, 2498; (f) C. Hilger and R. Stadler, *Macromolecules*, 1992, **25**, 6670; (g) M. Müller, U. Seidel and R. Stadler, *Polymer*, 1995, **36**, 3143.
24. C. P. Lillya, R. J. Baker, S. Hütte, H. H. Winter, Y.-G. Lin, J. Shi, L. C. Dickinson and J. C. W. Chien, *Macromolecules*, 1992, **25**, 2076.
25. S. Sivakova, D. A. Bohnsack, M. E. Mackay, P. Suwanmala and S. J. Rowan, *J. Am. Chem. Soc.*, 2005, **127**, 18202.
26. S. J. Rowan, P. Suwanmala and S. Sivakova, *J. Polym. Sci., Part A: Polym. Chem.*, 2003, **41**, 3589.
27. K. Chino and M. Ashiura, *Macromolecules*, 2001, **34**, 9201.
28. P. Cordier, F. Tournilhac, C. Soulié-Ziakovic and L. Leibler, *Nature*, 2008, **451**, 977.
29. D. Montarnal, P. Cordier, C. Soulié-Ziakovic, F. Tournilhac and L. Leibler, *J. Polym. Sci., A: Polym., Chem.*, 2008, **46**, 7925.
30. Y. Chen, A. M. Kushner, G. A. Williams and Z. Guan, *Nature Chem.*, 2012, **4**, 467.
31. Y. Yanagisawa, Y. Nan, K. Okuro and T. Aida, *Science*, 2017, **359**, 72.

32. (a) J. F. Stoddart, *Chem. Soc. Rev.*, 2009, **38**, 1802.; (b) C. Pezzato, C. Y. Cheng, J. F. Stoddart and R. D. Astumian, *Chem. Soc. Rev.*, 2017, **46**, 5491.
33. C. G. Claesens and J. F. Stoddart, *J. Phys. Org. Chem.*, 1997, **10**, 254.
34. D. A. Leigh, *Angew. Chem., Int. Ed.*, 2016, **55**, 14506.
35. (a) Z. Zhu, C. J. Cardin, Y. Gan and H. M. Colquhoun, *Nat. Chem.*, 2010, **2**, 653; (b) H. M. Colquhoun, Z. Zhu, C. J. Cardin, Y. Gan and M. G. B. Drew, *J. Am. Chem. Soc.*, 2007, **51**, 16163; (c) H. M. Colquhoun and Z. Zhu, *Angew. Chem., Int. Ed.*, 2004, **43**, 5040; (d) M. P. Parker, C. A. Murray, L. R. Hart, B. W. Greenland, W. Hayes, C. J. Cardin and H. M. Colquhoun, *Cryst. Growth. Des.*, 2018, **18**, 386; (e) J. S. Shaw, R. Vaiyapuri, M. P. Parker, C. A. Murray, K. J. Lim, C. Pan, M. Knappert, C. J. Cardin, B. W. Greenland, R. Grau-Crespo and H. M. Colquhoun, *Chem. Sci.*, 2018, **9**, 4052.
36. B. W. Greenland, S. Burattini, W. Hayes and H. M. Colquhoun, *Tetrahedron*, 2008, **64**, 8346.
37. S. Burattini, H. M. Colquhoun, B. W. Greenland and W. Hayes, *Faraday Discuss.*, 2009, **143**, 251.
38. S. Burattini, H. M. Colquhoun, J. D. Fox, D. Friedmann, B. W. Greenland, P. J. F. Harris, W. Hayes, M. E. Mackay and S. J. Rowan, *Chem. Commun.*, 2009, **44**, 6717.
39. S. Burattini, B. W. Greenland, W. Hayes, M. E. Mackay, S. J. Rowan and H. M. Colquhoun, *Chem. Mater.*, 2011, **23**, 6.
40. S. Burattini, B. W. Greenland, D. H. Merino, W. Weng, J. Seppala, H. M. Colquhoun, W. Hayes, M. E. Mackay, I. W. Hamley and S. J. Rowan, *J. Am. Chem. Soc.*, 2010, **132**, 12051.
41. L. R. Hart, J. H. Hunter, N. A. Nguyen, J. L. Harries, B. W. Greenland, M. E. Mackay, H. M. Colquhoun and W. Hayes, *Polym. Chem*, 2014, **5**, 3680.
42. For a historical review of halogen bonding see: G. Cavallo, P. Metrangolo, R. Mailani, T. Pilati, A. Priimagi, G. Resnati and G. Terraneo, *Chem. Rev.*, 2016, **116**, 2478.
43. L. C. Gilday, S. W. Robinson, T. A. Barendt, M. J. Langton, B. R. Mullaney and P. D. Beer, *Chem. Rev.*, 2015, **115**, 7118.
44. B. Mallada, A. Gallardo, M. Lamanec, B. de la Torre, V. Špirko, P. Hobza and P. Jelinek, 2021, **374**, 863.
45. T. Clark, M. Hennemann, J. S. Murrey and P. Politzer, *J. Mol. Model.*, 2007, **13**, 291.
46. R. Kampes, S. Zechel, M. D. Hager and U. S. Schubert, *Chem. Sci.*, 2021, **12**, 9275.
47. (a) R. Tepper, S. Bode, R. Geitner, M. Jäger, H. Görls, J. Vitz, B. Dietzek, M. Schmitt, J. Popp, M. D. Hager and U. S. Schubert, *Angew. Chem., Int. Ed.*, 2017, **56**, 4047; (b) J. Dahlke, R. Tepper, R. Geitner, S. Zechel, J. Vitz, R. Kampes, J. Popp, M. D. Hager and U. S. Schubert, *Polym. Chem.*, 2018, **9**, 2193.
48. S. Nomimura, M. Osaki, J. Park, R. Ikura, Y. Takashima, H. Yamaguchi and A. Harada, *Macromolecules*, 2019, **52**, 2659.

49. G. Whittell, M. D. Hager, U. S. Schubert and I. Manners, *Nat. Mater.*, 2011, **10**, 176.
50. M. Burnworth, L. Tang, J. R. Kumpfer, A. J. Duncan, F. L. Beyer, G. L. Fiore, S. J. Rowan and C. Weder, *Nature*, 2011, **472**, 334.
51. (a) K. Kunzelman, M. Kinami, B. R. Crenshaw, J. D. Protasiewicz and C. Weder, *Adv. Mater.*, 2008, **20**, 119; (b) B. Crenshaw, M. Burnworth, D. Khariwala, P. A. Hiltner, P. T. Mather, R. Simha and C. Weder, *Macromolecules*, 2007, **40**, 2400.
52. A. R. Hurst, B. Escuder, J. F. Miravet and D. K. Smith, *Angew. Chem., Int. Ed.*, 2008, **47**, 8002.
53. M. Nakahata, Y. Takashima, H. Yamaguchi and A. Harada, *Nat. Commun.*, 2011, **2**, 511.
54. A. Phadke, C. Zhang, B. Arman, C.-C. Hsu, R. A. Mashelkar, A. K. Lele, M. J. Tauber, G. Arya and S. Varghese, *Proc. Natl. Acad. Sci. U. S. A.*, 2012, **109**, 4383.
55. M. Zhang, D. Xu, X. Yan, J. Chen, S. Dong, B. Zheng and F. Huang, *Angew. Chem., Int. Ed.*, 2012, **51**, 7011.
56. K. Imato, M. Nishihara, T. Kanehara, Y. Amanoto, A. Takahara and H. Otsuka, *Angew. Chem., Int. Ed.*, 2012, **51**, 1138.
57. G. Deng, F. Li, H. Yu, F. Liu, C. Liu, W. Sun, H. Jiang and Y. Chen, *ACS Macro Lett.*, 2012, **1**, 275.
58. P. Sahoo, R. Sankolli, H.-Y. Lee, S. R. Raghavan and P. Dastidar, *Chem. – Eur. J.*, 2012, **18**, 8057.
59. G. R. Desiraju, *Angew. Chem., Int. Ed.*, 1995, **34**, 2311.
60. (a) R. Vaiypuri, B. W. Greenland, H. M. Colquhoun, J. M. Elliott and W. Hayes, *Polym. Int.*, 2014, **63**, 93; (b) R. V. S. P. Sanka, B. Krishnakumar, Y. Leterrier, S. Pandey, S. Rana and V. Michaud, *Front. Mater.*, 2019, **6**, 137; (c) V. K. Thakur and M. R. Kessler, *Polymer*, 2015, **69**, 369.
61. S. Parida, S. Kullu, S. Hota, S. Mishra, *Polymer Composites: Fundamentals and Applications*, 2024, Springer, Singapore, 2024, ISBN: 978-981-97-2074-3.
62. J. D. Fox, J. J. Wei, B. W. Greenland, S. Burattini, W. Hayes, H. M. Colquhoun, M. E. Mackay and S. J. Rowan, *J. Am. Chem. Soc.*, 2012, **134**, 5362.
63. For a review of CNC nanocomposites see: S. J. Eichorn, A. Dufresne, M. Aranguren, N. E. Marcovich, J. R. Capadona, S. J. Rowan, C. Weder, W. Thielemans, M. Roman, S. Renneckar, W. Gindl, S. Veigel, J. Keckes, H. Yano, K. Ade, M. Nogi, A. N. Nakagaito, A. Mangalam, J. Simonsen, A. S. Benight, A. Bismarck, L. A. Berglund and T. Peijis, *J. Mater. Sci.*, 2010, **45**, 1.
64. N. Ouali, J. Y. Cavaille and J. Perez, *Plast. Rubber, Compos. Process. Appl.*, 1991, **16**, 55.
65. (a) R. Vaiyapuri, B. W. Greenland, S. J. Rowan, H. M. Colquhoun, J. M. Elliott and W. Hayes, *Macromolecules*, 2012, **45**, 5567; (b) R. Vaiyapuri,

B. W. Greenland, H. M. Colquhoun, J. M. Elliott and W. Hayes, *Polym. Chem.*, 2013, **4**, 4902.
66. M. V. Biyani, E. J. Foster and C. Weder, *ACS Macro Lett.*, 2013, **2**, 236.
67. Q. Zheng, Z. Ma and S. Gong, *J. Mater. Chem. A*, 2016, **4**, 3324.
68. A. Okada and W. Oppermann, *Macromol. Mater. Eng.*, 2006, **291**, 1449.
69. (a) K. Haraguchi and T. Takehisa, *Adv. Mater.*, 2002, **14**, 1120; (b) K. Haraguchi, T. Takehisa and S. Fan, *Macromolecules*, 2002, **35**, 10162; (c) K. Haraguchi, R. Farnworth, A. Ohbayashi and T. Takehisa, *Macromolecules*, 2003, **36**, 5732; (d) K. Haraguchi, H. J. Li, K. Matsuda, T. Takehisa and E. Elliott, *Macromolecules*, 2005, **38**, 3482.
70. Q. Wang, J. L. Mynar, M. Yoshida, E. Lee, M. Lee, K. Okuro, K. Kinbara and T. Aida, *Nature*, 2010, **463**, 339.
71. K. Haraguchi, K. Uyama and H. Tanimoto, *Macromol., Rapid Commun.*, 2011, **32**, 1253.
72. *Handbook of Adhesive Technology*, ed. A. Pizzi and K. L. Mittal, CRC Press, Boca Raton, 3rd edn, 2017.
73. (a) A. H. Hofman, I. A. van Hees, J. Yang and M. Kamperman, *Adv. Mater.*, 2018, **30**, 1704640; (b) C. Heinzmann, C. Weder and L. M. De Espinosa, *Chem. Soc. Rev.*, 2016, **45**, 342.
74. K. Yamauchi, J. R. Lizotte and T. E. Long, *Macromolecules*, 2003 **36**, 1083.
75. A. D. O'Donnell, S. Salimi, L. R. Hart, T. S. Babra, B. W. Greenland and W. Hayes, *React. Funct. Polym.*, 2022, **172**, 105209.
76. G. M. L. Van Gemert, J. W. Peeters, S. H. M. Sontjens, H. M. Janssen and A. W. Bosman, *Macromol. Chem. Phys.*, 2012, **213**, 234.
77. (a) T. S. Babra, A. Trivedi, C. N. Warriner, N. Bazin, D. Castiglione, C. Sivour, W. Hayes and B. W. Greenland, *Polym. Chem.*, 2017, **8**, 7207; (b) T. S. Babra, M. Wood, J. S. Godleman, S. Salimi, C. Warriner, N. Bazin, C. R. Siviour, I. W. Hamley, W. Hayes and B. W. Greenland, *Eur. Polym. J.*, 2019, **119**, 260.
78. T. S. Babra, C. Warriner, N. Bazin, W. Hayes and B. W. Greenland, *Mater. Today Commun.*, 2021, **26**, 101777.
79. C. Heinzmann, S. Coulibaly, A. Roulin, G. L. Fiore and C. Weder, *ACS Appl. Mater. Interfaces*, 2014, **6**, 4713.
80. (a) D. H. Bae, P. H. Shin, S. T. Kwak, M. J. Moon, M. Y. Shon, S. T. Oh and G. N. Kim, *J. Ind. Eng. Chem.*, 2015, **30**, 92; (b) Patent: Reiss E. A., Malofsky A. G., Barber J. P. and Bauer D. P., Method of Adhesive Bonding by Induction Heating, *US* 6639197 B2, 2003; (c) Patent: Novorsky D. E., Bednarz R. J., Induction Heating for Adhesive Bonding, 4749833, 1988.
81. (a) S. Salimi, T. S. Babra, G. S. Dines, S. W. Baskerville, W. Hayes and B. W. Greenland, *Eur. Polym. J.*, 2019, **121**, 109264; (b) S. H. Vattathurvalappil and M. Haq, *Composites, Part B*, 2019, **175**, 107162; (c) X. Cheng, Y. Zhou, A. D. M. Charles, Y. Yu, M. S. Islam, S. Peng, J. Wang, A. N. Rider, M. Lim, V. Timchenko and C.-H. Wang, *Mater. Des.*, 2021, **210**, 110076.

82. (a) A. Marinow, Z. Katcharava and W. H. Binder, *Polymers*, 2023, **15**, 1145; (b) W. Mai, Q. Yu, C. Han, F. Kang and B. Li, *Adv. Funct. Mater.*, 2020, **30**, 1909912.
83. C. Wang, H. Wu, Z. Chen, M. T. McDowell, Y. Cui and Z. Bao, *Nat. Chem*, 2012, **5**, 1042.
84. Y. Huang, J. Liu, J. Wang, M. Hu, F. Mo, G. Liang and C. Zhi, *Angew. Chem., Int. Ed.*, 2018, **130**, 9958.
85. A. Feula, X. Tang, I. Giannakopoulos, A. M. Chippindale, I. W. Hamley, F. Greco, C. P. Buckley, C. R. Siviour and W. Hayes, *Chem. Sci*, 2016, **7**, 4291.
86. For extensive studies of the structure and bonding in this class of PU see: (a) P. J. Woodward, D. Hermida-Merino, B. W. Greenland, I. W. Hamley, Z. Light, A. T. Slark and W. Hayes, *Macromolecules*, 2010, **43**, 2512; (b) D. Hermida-Merino, M. Belal, B. W. Greenland, P. Woodward, A. T. Slark, F. J. Davis, G. R. Mitchell, I. W. Hamley and W. Hayes, *Euro. Polym. J.*, 2012, **45**, 1249; (c) D. Hermida-Merino, A. Feula, K. Melia, A. T. Slark, I. Giannakopoulos, C. R. Siviour, P. Buckley, B. W. Greenland, D. Liu, Y. Gan, P. J. Harris, A. M. Chippindale, I. W. Hamley and W. Hayes, *Polymer*, 2016, **107**, 368.
87. For a review of 'electronic' skin see: M. L. Hammock, A. Chortos, B. C.-K. Tee, J. B.-H. Tok and Z. Bao, *Adv. Mater.*, 2013, **25**, 5997.
88. J. Chen, J. Liu, T. Thundat and H. Zeng, *ACS Appl. Mater. Interfaces*, 2019, **11**, 18720.

CHAPTER 5

Thermodynamics of Self-healing and their Application in Polymeric Materials

JIAHUI LIU,[a] YING YANG*[b] AND MAREK W. URBAN*[a]

[a] Department of Materials Science and Engineering, Clemson University, Clemson SC 29634, USA; [b] Department of Chemistry, University of Nevada, Reno, USA
*Emails: mareku@clemson.edu; yingy@unr.edu

5.1 Introduction

The target of producing man-made materials in general, and polymers in particular, capable of repeatedly repairing themselves is driven by a desire to create sustainable and environmentally compliant technologies.[1–3] Often inspired by biological systems, the self-healing of polymers is being viewed as perhaps one of the most attractive and technologically advanced features of future materials, particularly when repeatability and reversibility of the damage-self-healing cycle can be achieved. The fundamental difference between biological and synthetic materials is the ability of the former to metabolize, that is to effectively utilize or remove side products produced during self-healing cycles. Since synthetic materials do not exhibit these attributes, the design of efficient self-healing polymers is challenging and requires an interplay of physico-chemical processes at molecular and macroscopic scales which may or may not be reversible.

Intuitively, self-healing involves conformational changes of macromolecular segments that parallel reforming chemical reactions leading to network remodeling. The complexity of these transient events is

Polymer Chemistry Series No. 41
Healable Polymer Systems: Fundamentals, Synthesis and Applications, 2nd Edition
Edited by Barnaby W. Greenland and Wayne Hayes

Published by the Royal Society of Chemistry, www.rsc.org

compounded by numerous covalent and non-covalent bond reformations that may or may not be reversible. Furthermore, these reformation reactions often parallel each other with conformational and configurational changes of macromolecular segments with variable flexibilities. Thus, thermodynamic states of macromolecular chains and chemical reformability of individual segments will determine the success of self-healing reactions. The first part of this chapter focuses on thermodynamic aspects of self-healing and the role of entropic and enthalpic components of the Gibbs free energy in the context of recoupling the self-healing lattice model that accounts for the reactivity and flexibility of macromolecular segments. The second section of this chapter outlines recent notable advances of chemistries utilized in self-healing of thermoplastic and thermosetting polymers.

5.2 Thermodynamics of Self-healing

Initial studies on thermoplastic polymers suggested that crack healing involves five stages: segmental surface rearrangements, surface approach, wetting, diffusion, and randomization.[4] In essence, the outcome of this process was to regain mechanical properties at the repaired area through forming new chain entanglements. Thus, chain diffusion at the polymer–polymer interface was considered the primary driving force for repairs, and the reptation model[5] semi-quantified these events by describing the motion of macromolecular segments trapped in a stationary tube. Under these boundary conditions, self-healing was achieved only when the entire chain completely disengaged from the tube, reaching an equilibrium state after time T_r (T_r is the time required for complete chain disengagement from the tube). Using this concept, the motion of one chain end can be characterized by the relaxation of the chain end using Gaussian partition function Z (eqn (5.1)),[6]

$$Z(\xi_1, \xi_2) = (\gamma/\pi)^{1/2}\left[\exp\left\{-\gamma(\xi_1+\xi_2)^2\right\}+\exp\left\{-\gamma(\xi_1-\xi_2)^2\right\}\right] \tag{5.1}$$

where: ξ is the distance of an end segment from the interface, $\xi = \xi_1$ and $\xi = \xi_2$ represent the positions of the two ends of one chain, $\gamma = 1/2a^2N$, where a is the length of one segment, and N is the number of segments of a chain. When the equilibrium conformation is achieved, chains become Gaussian in nature, and Z can be expressed as shown in eqn (5.2):

$$Z_e(\xi_1, \xi_2) = (\gamma/\pi)^{1/2}\exp\left\{-\gamma(\xi_1-\xi_2)^2\right\} \tag{5.2}$$

Under these conditions, self-repairing process can be viewed as chain conformational changes going from non-Gaussian to equilibrated Gaussian state during which each chain escapes from the tube in a given time t, and its chain length (l) is given by eqn (5.3):

$$\langle l\rangle \approx 2(Dt/p)^{1/2}\left[1-\exp\left\{-L^2/2Dt\right\}\right] \tag{5.3}$$

The term $\langle l \rangle \pi t^{1/2}$ (eqn 5.2) represents early stages of self-healing and for short times t, the exponential term is small, but to complete the repair, each chain must diffuse out of the tube. Considering molecular weight (M) dependence,[4] the repair time is proportional to the cube of the molecular weight ($T_r \propto M^3$), thus suggesting that lower molecular weight polymers exhibit favorable repairing conditions. Assuming bond cleavage does not occur during damage, chain diffusion is the only driving force for repair, and the source of loose chain ends is chain slippage. The healing efficiency defined by the extent of the recovery with respect to its initial post-damage state can then be expressed as the ratio of stress $R(\sigma)$ before and after healing (eqn (5.4)):

$$R(\sigma) = \frac{\sigma_{\text{healed}}}{\sigma_{\text{initial}}} \tag{5.4}$$

It has also been shown that the fracture stress, σ, is directly proportional to $(t/M)^{1/4}$.[6] Mathematically, if repair is considered as the rearrangement of chain conformations from non-Gaussian to Gaussian states, the time dependent evolution of chain lengths as well as the fracture stresses will again favor low molecular weight polymers that exhibit shorter repairing times and energetically favorable stress recoveries. These chains also exhibit lower glass transition temperatures (T_g), in particular near the interfacial regions. However, chain slippage will not be the sole source of active chains. During repair, physical or chemical damage often will cause chain ruptures, resulting in reactive chain ends that, as will be seen, significantly contribute to self-healing.

While prior studies were useful for when diffusion is the primary driving force for self-healing,[4] self-repairing involves other physico–chemical interactions at various time scales. Recent studies have shown that favorable molecular interactions within a polymer matrix will facilitate self-healing *via* supramolecular chemistry[7] or multi-level covalent bond reformation.[8] As damage occurs, chains cleave at the weakest points, such as hydrogen bonding, metal–ligand coordination sites, or the weakest covalent bonds. Although covalent bond cleavage may require higher energies, they typically generate free radicals and their stability is critical to achieve desirable repairs. Regardless of the energy input and chemistries involved, mobile and reactive chain ends capable of bond reforming will exhibit higher kinetic and potential energies. At the same time, release of constraints in localized areas from the surrounding polymer matrix will be manifested by relaxations of loose chain entanglements at newly created interfaces. This is schematically depicted in Figure 5.1(A).

After damage, being a non-equilibrium state, self-healing will result in a new Gaussian equilibrium state *via* localized reactions and associated chain relaxations primarily at loose chain ends. Energetically, self-repair will occur when the Gibbs free energy changes ($\Delta G = \Delta H - T\Delta S$) are <0. In the early stages, entropic contributions are the primary driving force, as soon as the

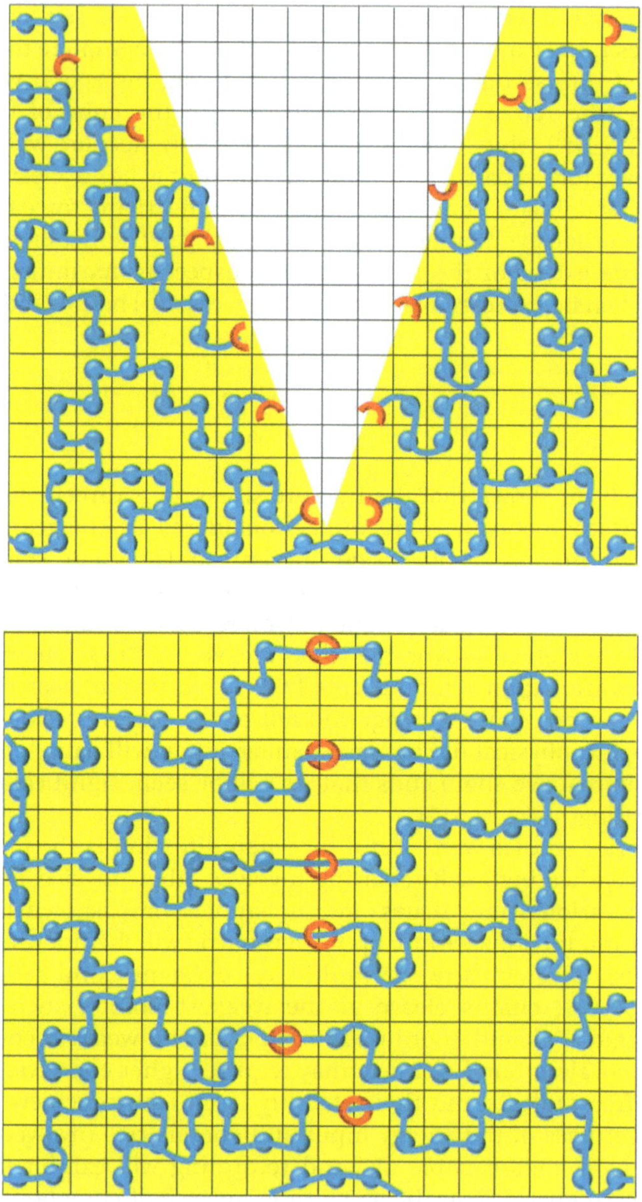

A

B

Figure 5.1 Recoupling self-healing lattice model of free chain ends at damaged polymer interfaces. A represents the damaged state with dangling chains and voids, and B is the repaired state with free chain ends crosslinked with their counterparts. Each square represents a single lattice site occupied by one segment of a polymer chain. Open circles represent reactive chain ends, whereas full circles are the remaining segments of the lattice.

chain ends react and approach the Gaussian state, the enthalpic term will dominate ΔG. This is schematically shown in Figure 5.1(B).

The following section will consider the role of enthalpic and entropic contributions showing that they are significantly different at different stages of self-healing for thermosetting and thermoplastic polymers and that entropic contributions will play a dominant role for higher disordered networks.

Assuming that all the chains are energetically equal and there is no energy exchange between neighboring chains, the entropy change for a finite change from State 1 (Figure 5.1(A)) to State 2 (Figure 5.1(B)) during repair can be expressed as eqn (5.5):

$$\Delta S = S_2 - S_1 = k\ln(\Omega_2/\Omega_1) \tag{5.5}$$

where, k is the Boltzmann constant, and Ω_1 and Ω_2 are the numbers of available arrangements of chain segments at State 1 and State 2. Analogous to the spontaneous expansion of gas molecules in an open space, the volume increase near created interfaces will lead to an increase in the number of arrangements available for a given state. Therefore, driven by the positive entropy changes, chain ends will continuously expand into voids, particularly when two surfaces are favorably distant. However, unlike freely diffusing small molecules, polymeric chains exhibit limited mobility under the surroundings of polymer matrix confinements.

In view of these considerations, it can be assumed that there are two types of macromolecular domains represented as self-healing recoupling lattices generated as a result of mechanical damage: (1) free chains resulting from chain rupture with both chain ends being free, and (2) tethered chains having one end anchored to the surface and the other free dangling end, as depicted in Figure 5.1. Assuming that free chains exhibit both translational and configurational freedom, the latter aspect of tethered chains will be the primary contributor to the entropy changes. Furthermore, the confinements of chains near the anchored surface will result in smaller entropy changes, but will tend to increase toward their dangling ends. Since entropy changes as well as self-repair attributes are affected by the flexibility of chains, the influence of chain rigidity reflected by the glass transition temperature of a given polymer will have a significant effect on the number of chain configurations defined in eqn (5.1). Depicted in Figure 5.1(A) and (B) is a self-healing recoupling lattice model, with States 1 and 2 immediately after damage (Figure 5.1(A)) and during repair (Figure 5.1(B)), respectively. As shown in Figure 5.1, for n self-avoiding walk polymer chains, each has N segments (a segment is defined as a repeating unit or a macromolecular unit that maintains its conformation during repair), (N-1) bonds and (N-2) bond angles, fitted into a self-healing recoupling lattice, where each segment occupies one lattice site. The entropy changes will be principally localized near the damaged area primarily occupied by dangling ends (half circles, Figure 5.1). If one considers chains being semi-flexible, the flexibility parameter, f, defined as the probability of two successive bonds to be not collinear and are subject to rotations, so that out of $n(N\text{-}2)$ bonds, there will be $fn(N\text{-}2)$ bent bonds. Under the condition that $\partial\Delta G/\partial f = 0$, f can be given as eqn (5.6):[9]

$$f = \frac{(z-2)e^{-\varepsilon/k_B T}}{1+(z-2)e^{-\varepsilon/k_B T}} \tag{5.6}$$

where, z is the coordination number of the lattice unit, k_B is the Boltzmann constant, T is the absolute temperature, and ε is the energy difference between collinear and bent bonds. The primary consequence of mechanical perturbations of the lattice will be lowering molecular weight and removal of a fraction of polymer lattice sites, which will parallel the formation of new sites. The total number of chain conformations at the damaged site can be expressed as eqn (5.7).[9]

$$\Omega = \frac{1}{n!2^n}\begin{pmatrix}(N-2)n \\ f(N-2)n\end{pmatrix}\prod_{j=1}^{n} \nu_j \tag{5.7}$$

where, ν_j accounts for the total number of chain end configurations. Since eqn (5.7) provides the binomial distribution of chain end configurations that account for the probability of finding $fn(N\text{-}2)$ bent bonds out of a total of $n(N\text{-}2)$ bonds (assuming that the energy of molecular interactions is zero), ΔG for free chains with two ends free can be given as eqn (5.8):[9]

$$\Delta G = -T\Delta S = -nRT\{\ln N + \ln(z/2e) + (N-2)\ln[1/(1-f)e]\} \tag{5.8}$$

The first term, ln N, originates from the opportunity for the first chain segment to be in a random location. For tethered chains, assuming that the first segment remains in a fixed position during repair, $N = 1$ and $\ln N = 0$, and the remaining terms of eqn (5.8) will apply, assuming that ε is the same for all bonds (although ε may be larger at the tethered end than at the free chain end). Thus, ΔG for tethered chains becomes eqn (5.9):

$$\Delta G = -T\Delta S = -nRT\{\ln(z/2e) + (N-2)\ln[1/(1-f)e]\} \tag{5.9}$$

According to eqn (5.8) and (5.9), f will have the threshold value K, such that when $f > K$, $\Delta G < 0$ and the chains will be able to move spontaneously. When $f < K$, the chains will be too rigid to move, making repair unlikely to occur, or requiring higher external energy inputs.

For free chains, it can be shown from eqn (5.8) that:

$$K_F = 1 - (1/e)(zN/2e)^{1/N-2} \tag{5.10}$$

whereas for tethered chains, K becomes eqn (5.11):

$$K_T = 1 - (1/e)(z/2e)^{1/N-2} \tag{5.11}$$

Assuming that tethered chains have the coordination number $z = 6$ that consist of 10 segments, $K_T = 0.63$, and $f > 0.63$ for self-healing to occur. For a free chain with 10 segments, however, $K_F = 0.52$, implying that to initiate conformational changes, lower flexibility will be necessary. Since the K value increases with the increasing number of chain segments, therefore, longer chains will require higher flexibility to acquire movement, but the difference for the chains with 1000 and 10 segments can be as small as 0.01.

According to eqn (5.6), at $T = \text{constant}$, f will be determined by the energy difference between bent and collinear bond states ε. As the self-repairing process depicted by the recoupling self-healing lattice model in Figure 5.1 continues, chain ends come into contact with each other to form new cross-linked sites or entanglements, leading to the recovery of mechanical properties.

Free chains exhibit higher f values as a result of lower ε, and typically have smaller N when compared to tethered chain populations, which result in more negative ΔG and greater mobility. When free chains react with each other and add onto tethered chains, they contribute to the increase of N. As this process continues, chains become more rigid and, as N increases until f reaches the threshold value of K for a given N, ΔG will be positive. At this point, a stable Gaussian equilibrium state is reached. When $f < K$ after damage, repair will not occur. According to eqn (5.9), low molecular weights as well as low T_g (low ε, high f) polymers will favorably contribute to higher ΔS values, thus resulting in $\Delta G < 0$. These attributes will facilitate chain end motions toward repair, and are one of the prerequisites, but not sole requirements for self-healing to occur. Self-repair *via* diffusion may not be achieved when enthalpic components dominate macromolecular miscibility unless inter- and intra-molecular interactions defined by the Flory–Huggins parameter χ are favorable or an external energy source is applied.

Considering internal energy contributions to self-repairing, the internal energy changes $\Delta U = \Delta Q - W$ (where, U is the internal energy, ΔQ is the heat, and W is the work done to the surroundings or any other form of external energy input). Since enthalpy $H = U + pV$ (p is pressure, V is volume). At $p = \text{constant}$, $\Delta U = \Delta Q - P\Delta V + E_{ext}$ (where, E_{ext} is any other form of external energy input to the system except Q and W). This may be expressed as in eqn (5.12):

$$\Delta H = \Delta Q + E_{ext} \tag{5.12}$$

Assuming that $p = \text{constant}$ during repair, ΔH is the sum of heat transfer Q associated with chain rearrangements and chain ends reactions at the interface as well as any other form of energy input from damaged to repaired states. In addition to entanglements of chain ends at two surfaces, reactive counterparts at each chain end, such as non-hydrogen bonded ends, non-coordinated sites, or reactive free radicals, will contribute to ΔH. These reactive macromolecular segments generated by damage should retain their reactivity in order to participate in self-healing reactions. Thus, interactions

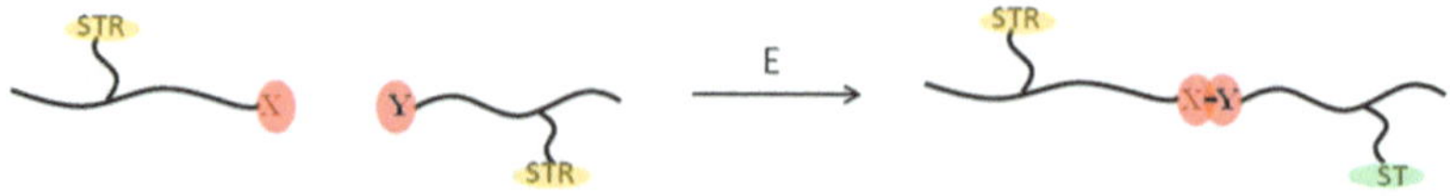

Scheme 5.1 Schematic representation of cleavage and reformation of covalent bonds in macromolecules that contain stimuli-responsive component (STR).

with neighbors from the same interface (Figure 5.1(A)), or quenched by any other gas phase molecules, will not favor self-healing reactions.

Energetic contributions to ΔH are critical and, at the onset of repair, may need to be triggered by built-in stimuli-responsive components. For example, external stimuli can be utilized to break certain linkages which may enhance mobility or create reactive sites for future reactions. Ideally, these stimuli network components need to be able to reform after repair, otherwise repair may not be reversible. Scheme 5.1 illustrates the reaction of two macromolecular segments that contain a stimuli-responsive component (STR) being able to regenerate itself, and reactive chain ends X and Y, whereas:

ST represents a fraction of STR that is unable to recover to its original state, and E represents an external energy input. The enthalpy change associated with the healing process can be approximated by bond energies as eqn (5.13):

$$\Delta H = [\omega_{X-Y} - (\omega_X + \omega_Y)] + [\omega_{ST} - \omega_{STR}] \tag{5.13}$$

where, ω is the sum of bond energies of each component. If the chain end groups react with neighbors from the same surface or quenched by other species, this may result in similar total enthalpy changes, but these conditions are undesirable for self-healing.

Damage–repair processes can be reversible or irreversible. When the stimuli-responsive component STR is present, STR absorbs energy in order to be activated to a higher energy state, and triggers the onset of repair. After removal of the stimuli, a fraction of STR may recover giving off heat. If STR is completely recovered, the process shown in Scheme 5.1 can be carried out again and be "reversible self-repairing". If STR cannot be recovered at all, only one time repair – "irreversible repair" is anticipated. STR may also be partially recovered, and a limited number of repair cycles is expected.

For irreversible infinitesimal irreversible self-healing processes to occur, $\Delta S > \Delta Q/T$. Substituting ΔQ with eqn (5.12), $E_{ext} > \Delta H - T\Delta S$, and the repair process will occur according to eqn (5.14).

$$\Delta G - E_{ext} < 0 \tag{5.14}$$

Most efficient self-healing can be achieved when their chain ends recouple ($\Delta H < 0$) exothermically with $\Delta S > 0$. One example of such networks is polymers with the ability of free radical recoupling or hydrogen bonding formation. At the same time, if chain mobility is high enough to facilitate diffusion, external energy input may not be necessary. When self-healing reactions are endothermic ($\Delta H > 0$), or a polymer network is too rigid to undergo conformational changes ($\Delta S < 0$), E_{ext} will be required to overcome the positive ΔG. As indicated earlier, E_{ext} may activate built-in stimuli-responsive components leading to an increased kinetic energy of the chains, thus decreasing the energy barrier for chain end recoupling reactions.

Although each polymer system will have its own characteristics, in general, for polymeric solids, especially polymer gels, entropic contributions favor self-repair. As chains undergo elastic deformations and inter-diffusion at localized areas, they will take new conformations and ultimately fill up the voids created by mechanical damage. If bond formation during repair is exothermic, it contributes to more negative ΔG. However, if the activation of stimuli-responsive components (STR) requires external energy input I, or reactive chain ends can be quenched to another ground state before repair, additional external energy input will be needed to overcome the new energy barriers. Realizing that each self-healing process involves chemical reactions and physical network rearrangements, self-healing will be specific to a given polymer matrix. Figure 5.2(A)–(C) illustrate the general energetic contributions of a stimuli-responsive behavior for solutions, gels, and solids.[10]

The transitions between these minima represent the energy required for a system to undergo transition from one state to another, which can be achieved easily in solutions when Brownian motion energies are low and conformational changes require small energy inputs. This is illustrated by two energy minima shown in Figure 5.2(A), which facilitate the equally easily achievable energy states in two different conformations. For gel networks with higher density, yet typically low T_g, one needs to take into account the stability of the network illustrated by the lower energy state ($\Delta E_{eq(Gel)}$) shown in Figure 5.2(B), which is associated with two smaller energy minima A and B representing the stimuli-responsive transition between STR and ST as a prerequisite for self-healing. When considering the solid state, the primary difference is the depth of the minimum energy of $E_{eq(Solid)}$ responsible for network stability. The latter are responsible for network repairing, whereas A (STR state) and B (ST state) will provide stimuli-responsiveness that enables self-healing.

The main question is how these energy distributions defined by the chemical makeup of the network can be utilized in self-healing processes. When damage occurs, the lowest energy state ($E_{eq(Solid)}$) will rapidly change to a less stable state and macromolecular segments will have high mobility, reflected by lower T_g values near the surface/interface created by damage. At the same time, energy contributions from self-healing elements will exhibit higher energy states in order to respond to damage and initiate the repair process. Upon repair, the $E_{eq(Solid)}$ of solids will be restored, but A and B may or may not. If self-healing is reversible, A and B will regenerate again to the lower energy levels at the expense of new equilibrium energy, but for irreversible self-healing, A minimum (STR state) will vanish.

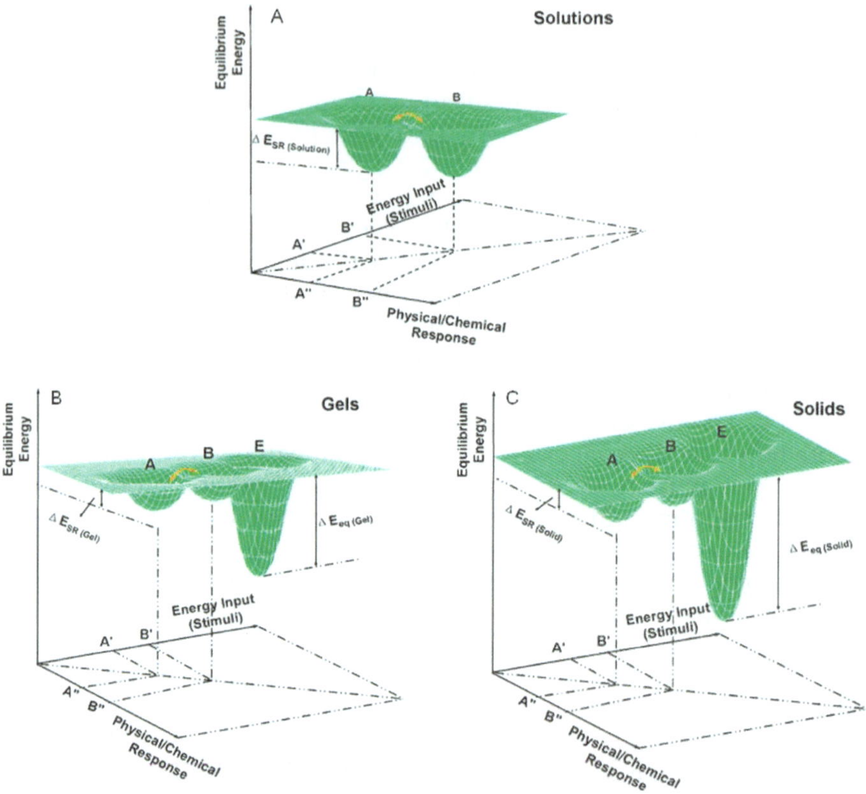

Figure 5.2 The relationship between equilibrium energy, stimuli energy input, and physical/chemical response: solutions (A), gels (B), and solids (C). Reproduced from ref. 10 with permission from Elsevier, Copyright 2010.

While these general considerations may serve as a guide for designing self-healable thermoplastic and thermosetting polymers, it should be realized that precise molecular design with controllable polymer topologies are necessary to meet thermodynamic requirements. The last decade offered a variety of chemical approaches offering self-healing in thermoplastic and thermosetting polymers of which some are reversible, and some are not.[11] The following section outlines recent advances in thermoplastic and thermosetting polymers that exhibit both reversible and irreversible characteristics.

5.3 Self-healing Reactions in Thermoplastic and Thermosetting Polymers

Regardless of the thermoplastic or thermosetting nature of a given polymer, the recoupling self-healing lattice model provides an estimate of which of the network components predominantly contributes to the transition between non-equilibrium to equilibrium states during self-repair. Parallel to conformational and configurational changes, chemical events that contribute to self-healing are:

1. bond cleavage, that will generate reactive bond-reformable groups, such as COOH, NH_2, OH, SH, C=O, or free radicals;
2. chain slippage, that will result in configurational and conformational changes;
3. combination of supramolecular networks with reforming interactions, for example hydrogen bonding, metal–ligand coordination bonds, host–guest interactions, ionic interactions, or π–π interactions.

5.3.1 Bond Cleavage Reactions in Thermoplastic and Thermosetting Polymers

Since free radicals or other reactive chain ends generated by cleavage reactions may be intercepted by oxidative or other processes, one of the key challenges is to facilitate the stability of reactive groups during the network repair. For example, regardless of molecular weight, polycarbonate (PC) cannot self-repair upon mechanical damage in ambient environments. However, PC synthesis capable of self-repairing by ester exchange reactions can be accomplished using a steam pressure at 120 °C.[12] As shown in Figure 5.3(A), hydrolysis of carbonate groups under steam pressure generates phenolic chain ends, which can undergo ester exchange catalyzed by $NaHCO_3$ to recombine chain ends. As illustrated in Figure 5.3(B), poly(phenylene-ether) (PPE) repair processes will work in a similar manner.[13–15]

Damage along PPE backbones produces stabilized oxygen radicals by complexation with $Cu^{(II)}$ which further inhibits oxidation, and the presence of hydrogen donors facilitates their conversion to phenol groups. Thus, phenolic chain ends are able to recombine through ether exchange reactions at elevated temperatures catalyzed by $Cu^{(II)}$. These examples illustrate that built-in stimuli-responsive components, which are carbonate groups

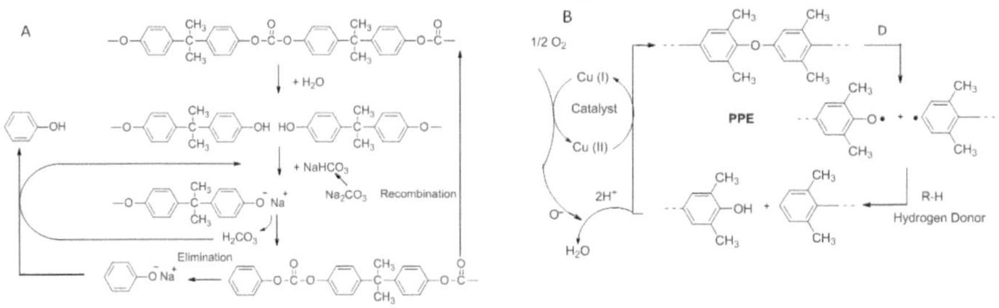

Figure 5.3 (A) Ester exchange reaction in polycarbonates leading to healing.[12] Reproduced from ref. 12 with permission from John Wiley & Sons, Copyright 2004 Wiley Periodicals, Inc. (B) Healing in polypropylene ethers *via* covalent bond formation between scission chain ends in the presence of a Cu catalyst.[13–15]

sensitive to pressured water steam, and hydrogen donors, respectively, enable recombination of chain ends leading to self-healing. It should also be be noted that PC and PPE are both high molecular weight rigid polymers, and ester and ether exchange reactions are endothermic in nature. Therefore, high temperatures are required to overcome the energy barriers of chain flexibility and elimination reactions in the presence of catalysts. These conditions meet energetic criteria related to ΔG and E_{ext} relationships discussed previously (eqn (5.14)). Furthermore, chain ends of these thermoplastic macromolecules are stabilized at the expense of consumption of stimuli-responsive reactants, and bond reforming reactions responsible for self-healing require elevated concentration levels of catalysts. One of the drawbacks resulting from the presence of high concentration levels of stimuli-responsive components are altered mechanical properties of the bulk material.

To minimize the influence of added components, cross-linked networks with reversible bonds were utilized to achieve both stable yet reactive chain end entities after damage as well as covalent bond-reformation enabling network repair. External energy input in the form of UV radiation and heating will activate the reversible reactions without chemical input, thus minimizing by-product formation. Figure 5.4(A) illustrates reactions of cinnamate monomer (1,1,1-tris-(cinnamoyloxymethyl)ethane (TCE)) cross-linked *via* [2+2] photocycloaddition that results in high-T_g transparent films.[16] Mechanical damage leads to the C–C bond cleavage of cyclobutane rings between TCE monomers resulting in the formation of original cinnamoyl groups, whereas the crack-healing characterized by recovery of flexural strength occurs by re-photocycloaddition to reform cyclobutane ring cross-links upon UV (>280 nm) exposure at 100 °C.[17–20]

Another approach illustrated in Figure 5.4(B) utilized the thermal reversibility of Diels–Alder (DA) and retro-DA reaction between furan–maleimide, dicyclopentadiene, and anthracene based polymers.[21–24] These reversible reactions were applied to epoxy, acrylate and polyamide systems, where retro-DA reactions result in disconnection of butylene oxide rings resulting in the formation of diene and dienophile. The reconstruction of covalent bonds repairing the crack occurs *via* recoupling of diene and dienophile

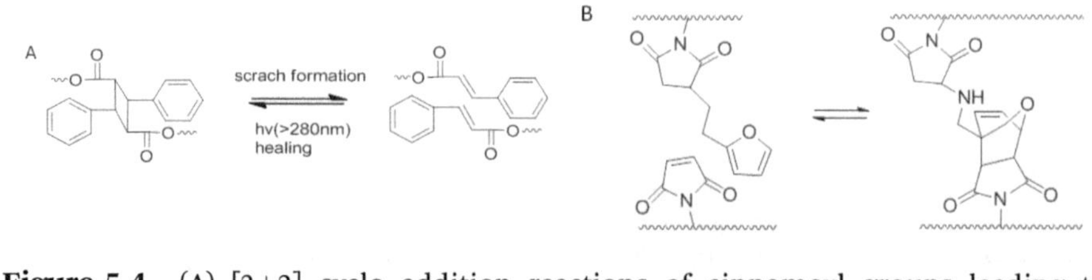

Figure 5.4 (A) [2+2] cyclo addition reactions of cinnamoyl groups leading to healing; (B) Diels–Alder mechanism of multifuran and multimalemide containing backbones within a cross-linked polymer system for reversible self-repairing.[13,20]

groups into butylene oxide rings. Again, reactive chain ends capable of re-forming linkages during repair are critical to the successful design of self-healing polymers.

As was demonstrated in Section 5.2, interfacial diffusion is one of the contributing elements to favorable Gibbs free energy changes during self-healing, and is enhanced in a melted state for polymers with high flexibilities, f_F. In poly(methyl methacrylate) (PMMA), the induced crack healing was achieved by heating it up to above the T_g under pressure.[25] To reduce T_g and facilitate healing, organic solvents are often used as plasticizers,[26,27] and the recovery of fracture stress was suggested to be proportional to $t^{1/4}$, thus agreeing with the predictions of the reptation model. Since localized elevated temperatures will enhance diffusion without solvent interference, self-healing thermoplastic polymers with uniformly dispersed γ-Fe_2O_3 nanoparticles were developed.[28] Prepared by *in situ* synthesis of *p*-methyl methacrylate/*n*-butylacrylate/heptadecaflurodecyl methacrylate (*p*-MMA/*n*BA/HDFMA) colloidal particles coalesce to form uniform films, which upon mechanical damage can be repaired by applying an oscillating magnetic field (OMF). Self-healing is achieved by activating oscillation of γ-Fe_2O_3 nanoparticles at the frequency of OMF, thus causing a localized temperature increase at the nanoparticle–polymer interface, generating interfacial melt flow to the damaged area and repair. In another approach, small grain thermoplastic epoxy particle adhesives were embedded within glass–epoxy composite matrices, which melted upon heating, thus repairing damages.[29] Although the voids left after melting of adhesives may have adverse effects on the integrity of the epoxy polymer matrix, this approach may serve many applications. Along the same lines, blending of thermoplastic poly(bisphenol-A-*co*-epichlorohydrin) with epoxy resin at 80 °C to dissolve it into the thermoset was utilized.[30] During damage, thermoplastic polymer in the vicinity of the crack flows to the newly created crack regions and heal the crack.

Ionic interactions in polymer matrices have been known to enhance mechanical properties and selected polymers have shown to exhibit self-healing attributes. In poly(ethylene-*co*-methacrylic acid) and polyethylene-*g*-poly(hexylmethacrylate), instead of external heat or an alternative stimulus (E_{ext}) to promote polymer diffusion in a damaged area, repairs were facilitated under ambient and elevated temperatures upon projectile puncture testing.[31,32] A ballistic puncture in low-density poly ethylene (LDPE) does not exhibit self-healing itself, whereas puncture in poly(ethylene-*co*-methacrylic acid) EMMA films heal the puncture leaving a scar on the surface. The proposed healing mechanism is a two-stage process in which upon projectile impact, the ionomeric network is disrupted, and heat generated by the friction during the damage is transferred to the polymer surroundings, generating a localized melt state. Elastic responses of the locally molten polymer facilitated by a puncture are known as puncture reversal. During the second stage, the molten polymer surfaces fuse together *via* interdiffusion resulting from ionic interactions between ionic clusters to seal the puncture, followed by rearrangement of the clustered regions and long-term relaxation processes, which continue until ambient temperatures are reached.

Combining repair and sensing attributes in one polymer matrix was accomplished in the colloidal synthesis of poly(methyl methacrylate/*n*-butylacrylate/2-[(1,3,3-trimethyl-1,3-dihydrospiro[indole-2,3′-naphtho[2,1-*b*][1,4]oxazin]-5-yl)amino]ethyl-2-methylacrylate) [p(MMA/*n*BA/SNO)], Figure 5.5(A). These films, upon mechanical damage, undergo color changes from clear to red in damaged areas. However, upon exposure to sunlight, temperature and/or acidic vapors, damaged areas are repaired and the initial colorless appearance is recovered. As shown in Figure 5.5(B). the process is reversible and driven by conformational changes of the statistical copolymer that parallels the ring opening–closure of spironapthoxazine (SNO) groups.[33] Upon mechanical damage, SNO segments of the neighboring segments form inter-molecular hydrogen bonding that stabilizes the copolymer backbone to reach an extended conformation which facilitates the expansion of polymer into the damaged area as a result of high mobility of the copolymer backbone. The following ring-closure of merocyanine (MC) to SNO resulting conformation change of polymer chains into a coiled state, leads to chain entanglements that facilitate the repair. This is illustrated in Figure 5.5(C).

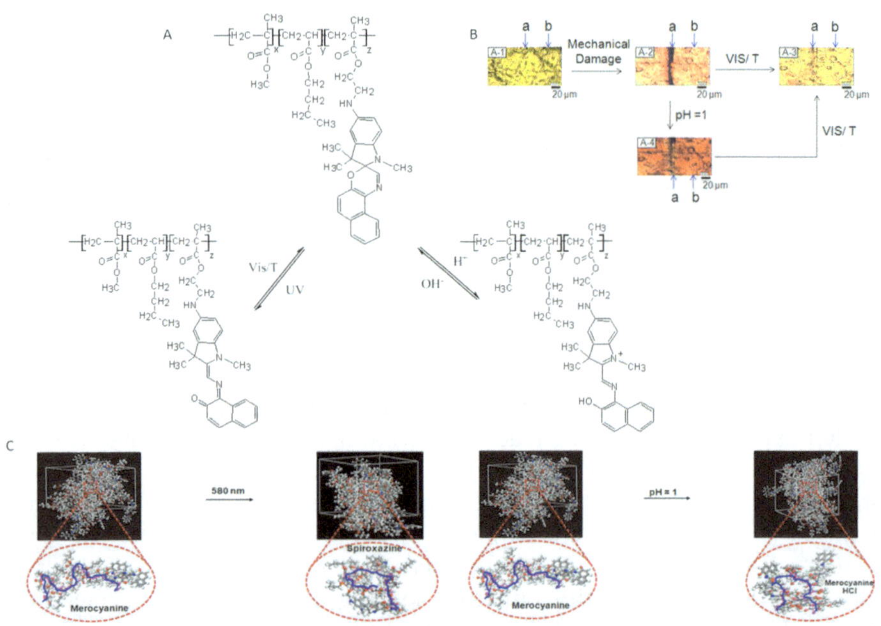

Figure 5.5 (A) Stimuli-responsiveness of the self-healing copolymer poly(methyl methacrylate/*n*-butylacrylate/2-[(1,3,3-trimethyl-1,3-dihydrospiro[indole-2,3′-naphtho[2,1-*b*][1,4]oxazin]-5-yl)amino]ethyl-2-methylacrylate) [p(MMA/*n*BA/ SNO)]; (B) optical images showing the color change upon scratching and self-healing behavior; (C) chain conformational changes facilitated by SNO ring-opening and closure. Adapted from ref. 33 with permission from the Royal Society of Chemistry.

These examples illustrate that self-healing in thermoplastics can be achieved by the introduction of localized molecular heterogeneities in the form of reactive groups or nano-insertions. In contrast, one of the challenges in designing thermosetting polymers is to incorporate self-healing components while maintaining useful mechanical properties often reflected by higher T_g. Crosslinked polyurethanes (PUR) represent these types of networks, but are unable to undergo self-healing on their own. However, when one of the most abundantly available natural polysaccharides, chitin, is modified by acetylation reactions to form chitosan (CHI), followed by further modification with the four member oxetane (OXE), and this macromonomer is cross-linked with tri-functional hexamethylene di-isocyanate (HDI) in the presence of polyethylene glycol, these PUR networks exhibit self-healing properties.[34] Upon mechanical damage of the network, self-repairing occurs upon exposure to UV radiation. The reactions leading to network formation are shown in Figure 5.6.

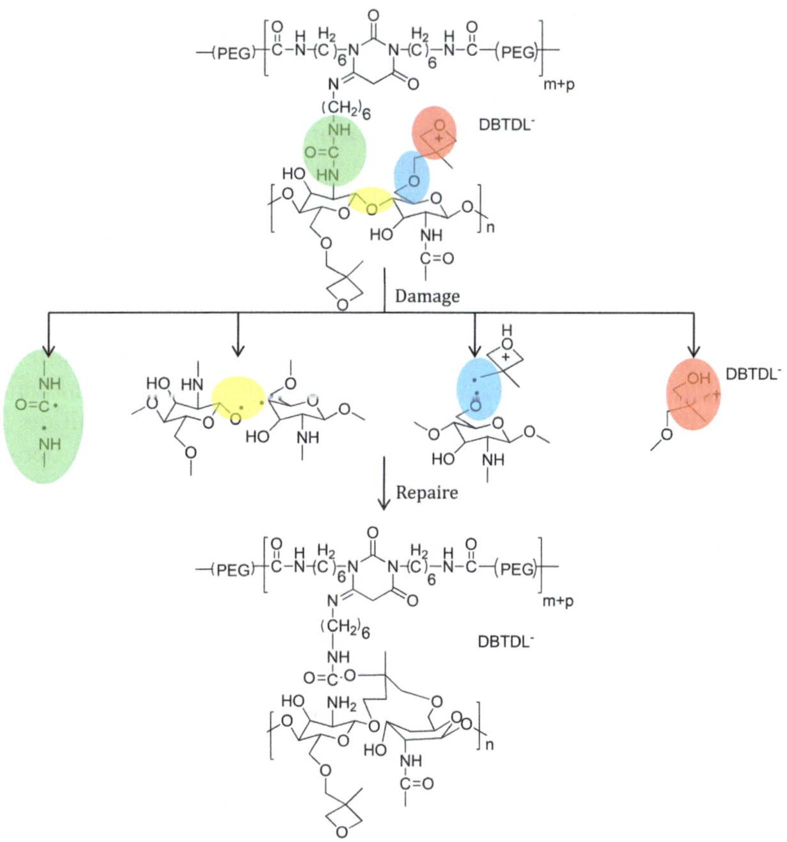

Figure 5.6 Synthetic steps involved in creating self-healing polyurethanes (PURs) containing oxetane (OXE) and chitosan (CHI). Adapted from ref. 35 with permission from the Royal Society of Chemistry.

PUR in combination with polyurea (PUA) generated as a result of a reaction between isocyanate and amine-functionalized OXE-CHI provides desirable integrity and localized heterogeneity of the network, but the OXE-CHI macromonomer facilitates cleavage of a constrained 4-member ring (OXE) to form free radicals and UV sensitivity (CHI) for self-repair.[35] Similar self-healing, although mechanistically quite different, was achieved for an oxolene-chitosan (OXO-CHI) macromonomer.[36] The choice of OXE and OXO rings was driven by the stability of free radicals generated by their respective ring-opening processes. The primary advantage of these systems is that the OXE-CHI and OXO-CHI macromonomers can be co-reacted to existing thermosetting polymers, ensuring ease of processability and desirable self-healing characteristics.

Dynamic covalent chemistry offers a unique alternative for self-healing of thermosetting polymers and gels capable of maintaining their mechanical integrity. As shown in Figure 5.7(A), trithiocarbonates (TTC) are able to undergo photostimulated reshuffling reactions in the same manner as photoinitiators in RAFT polymerization.[37–39] Therefore, a covalent cross-linked polymer matrix with enhanced segmental mobility was achieved using RAFT copolymerization of *n*-butyl acrylate (BA) and a TTC cross-linker. Upon mechanical damage, followed by UV exposure, repetitive network repairs were achieved, making this network a unique combination of dynamic covalent bond reshuffling reactions and network remodeling.

As a result of the reversible nature of disulfide (S–S) bonds achieved by reduction reactions to form thiol (S–H) groups which can be reversed by oxidation, S–S also offer a self-healing alternative.[40] These reactions are shown in Figure 5.7(B). Poly(*n*-butyl acrylate) grafted star polymers were prepared by atom transfer radical polymerization (ATRP) from cross-linked cores comprised of poly(ethylene glycol diacrylate) and further utilized as macroinitiators for the consecutive chain extension of bis(2-methacryloyloxyethyl disulfide) (DSDMA).[41] This approach introduces disulfide (S–S) reversible cross-links into the branch peripheries of cross-linked networks that upon cleavage under reducing conditions, form thiol (SH)-functionalized soluble star polymers. Grafted polymer architectures facilitate intrinsic low viscosity and high accessibility of functional groups that are critical in self-healing. Disulfide exchange reactions were also incorporated into low glass transition (T_g) gel networks to achieve room-temperature reversibility.[42]

Figure 5.7(C) illustrates the formation of diarylbibenzofuranone (DABBF) dynamic covalent bonds *via* stable arylbenzofuranone (ABF) radicals enabling reforming of DABBF at room temperature, thus facilitating self-healing.[43] These thermosetting hydrogel self-healable networks were prepared by reacting tetrahydroxy functionalized DABBF building blocks with a modified poly(propylene glycol) (PPG) bearing an isocyanate group at each end. As shown in Figure 5.7(C), when DABBF dissociates to give pairs of ABF radicals, it is critical that they exhibit little or no sensitivity to oxygen,[44] thus providing sufficient time for covalent bonds to reform during network remodeling,[1] otherwise oxidative processes will intercept the process.

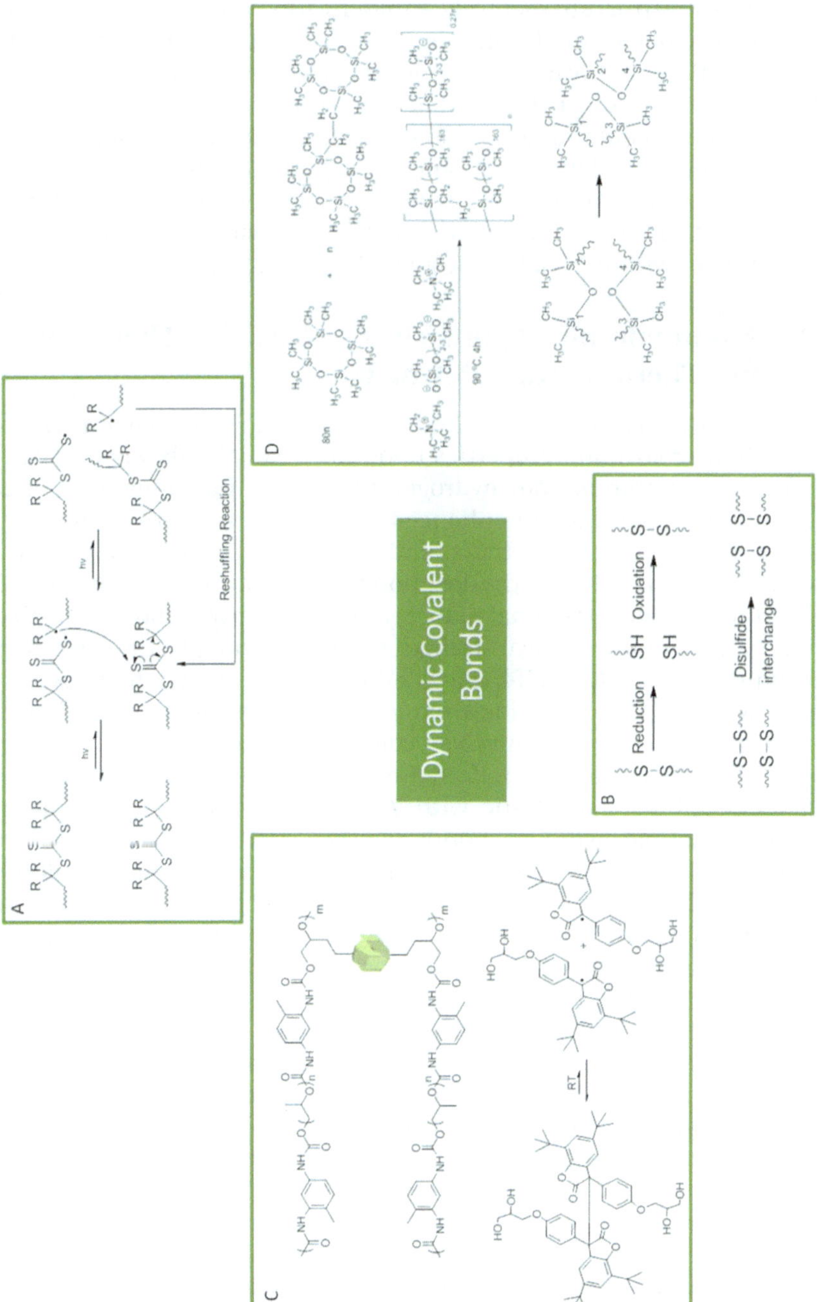

Figure 5.7 Schematic illustration of the mechanism of dynamic covalent bonds responsible for self-healing.[37,41,43,47]

Silicone-based polymers represent a broad spectrum of properties which can be either thermosets or thermoplastics. They are known for their ability to dynamically restructure themselves under various conditions,[45,46] and the recently revisited tetramethylammonium silanolate-initiated ring-opening copolymerization of octamethylcyclotetrasiloxane (D4) and bis(heptamethylcyclo-tetrasiloxanyl)ethane (bis-D4) showed again that these cross-linked polymers containing ethylene bridges and active silanolate end groups exhibit remodeling attributes.[47] Figure 5.7(D) depicts reactions leading to the formation of "living" tetramethylammonium dimethyl silamoate end groups, which maintain their activity under ambient conditions, thus providing "ready-to-respond" active groups.

5.3.2 Supramolecular Approaches to Healable Thermoplastic and Thermosetting Polymers

There are numerous studies that show how supramolecular chemistries may facilitate bond reforming properties upon physical stimuli. Figure 5.8(A) illustrates how combining four hydrogen bonds in a functional unit of urea isopyrimidone (Upy) leads to enhanced association strengths between Upy units and polysiloxane, polyethers, and polyesters.[48–50] Although relatively weak, hydrogen bonding is the key bond reforming ingredient, and as a result of their highly directional nature, multiple hydrogen bonds are able to facilitate sufficient mechanical strength to form thermoplastic elastomers.[51] Based on this concept, a self-healing rubber was synthesized *via* two steps using fatty diacids and triacids from renewable resources.[52] As shown in Figure 5.8(B), the first step involves condensation of acid groups with an excess of diethylenetriamine, followed by reactions with urea, and obtained a non-tacky rubber like plastic with $T_g = 8$ °C by adding dodecane as a plasticizer. Multiple hydrogen bonds between N–H and C=O groups are formed within the network, while crystallization is avoided. As a result of the advantage of high segmental mobility in hydrogels, other soft materials capable of self-healing utilizing hydrogen bonding with temperature or pH dependence have been developed.[53]

Copolymers provide the opportunity for designing multi-phase heterogeneous materials offering a broad range of mechanical properties. Figure 5.8(C) shows how hydrogen bonded brush polymers are able to self-assemble into a two-phase morphology, in which polystyrene backbone acts as the hard domain, in which soft-brushes are able to self-assemble *via* hydrogen bonding into soft-domains and behave like thermoplastic elastomers.[54] These materials are able to self-repair by bringing two cut ends together at room temperature without external heat and forming hydrogen bonding between –C=O groups and amide or amine-functionalized ends. In a thermodynamic sense, a self-healing temperature above T_g will ensure sufficient chain mobility to achieve $\Delta S > 0$. Since bond reforming using hydrogen bonding interactions is an exothermic process, an overall $\Delta G < 0$ will facilitate self-healing at room temperature without external stimuli (E_{ext}).

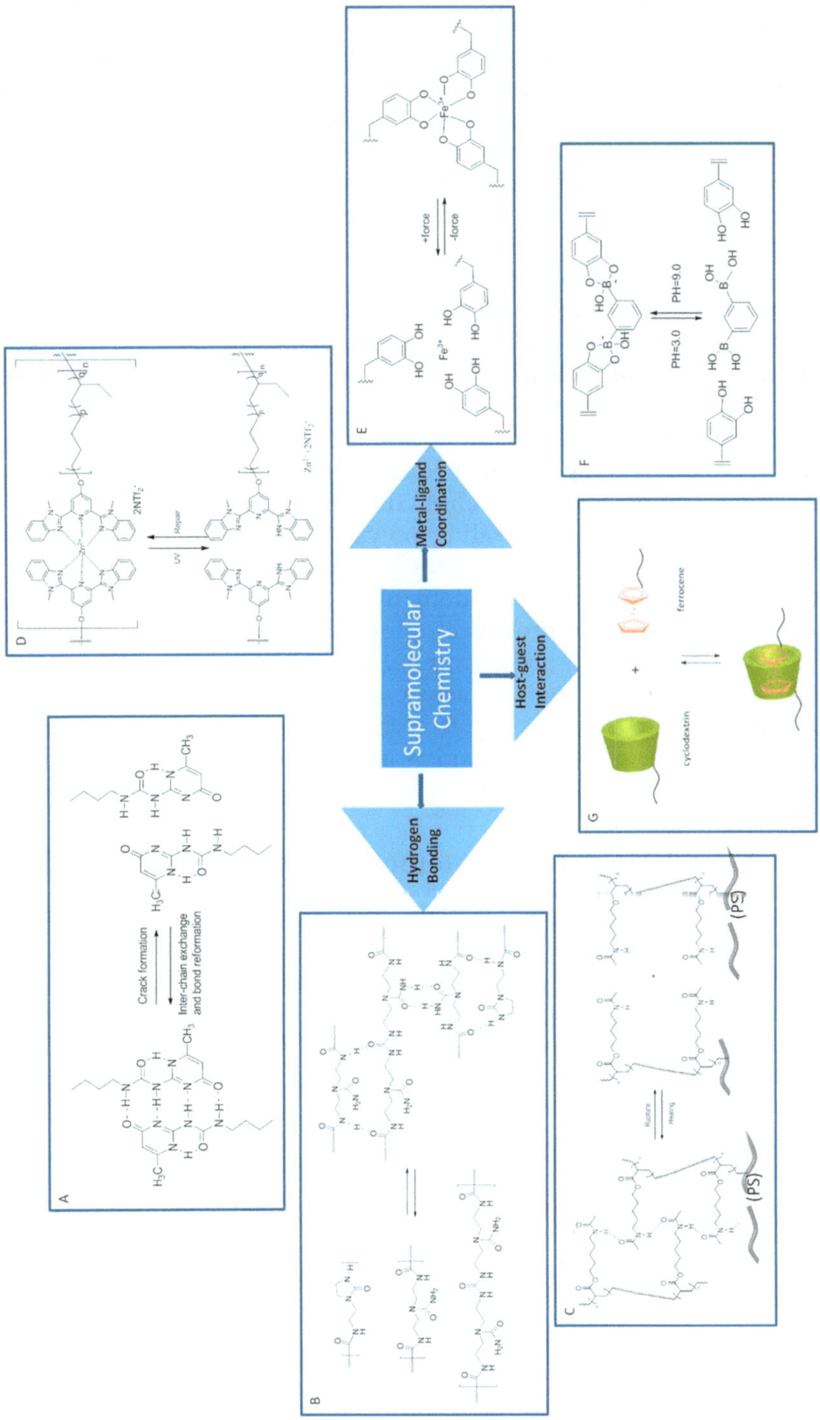

Figure 5.8 Self-healing model of supramolecular interactions including hydrogen bonding, metal–ligand coordination, and guest–host interaction.[48,52,54,57,59–61]

In a manner similar to hydrogen bonding, metal–ligand supramolecular interactions are effective in the design of supramolecular healable polymers.[55,56] As a result of their optical and photophysical properties, metal complexes offer many advantages and the reversibility of reactions can be tuned by different metal ion and ligand substitutes. Figure 5.8(D) shows phase separated supramolecular polymers based on a macromonomer comprising of a hydrophobic poly(ethylene-*co*-butylene) core with 2,6-bis(1′-methylbenzimidazolyl)pyridine ligands at the termini and were assembled by complexation with Zn^{2+}.[57] Upon exposure to UV light, the metal–ligand motifs are electronically excited and the absorbed energy is converted into heat. At the same time, elevated temperatures cause disengagement of metal–ligand bonds resulting in a decrease in molecular weight and viscosity. Facilitated by high temperatures produced by UV radiation, a favorable ΔS is obtained so that lower molecular weight chains are able to diffuse into each other, reforming metal–ligand complexation upon cooling. Recent studies also took advantage of the coordination between Fe and catechol ligands, as well as boronate–catechol complexation, which resulted in a pH-induced cross-linked self-healing polymer with near-covalent elastic moduli.[58–60] As depicted in Figure 5.8(E) and (F), the complexation ability between metal–catechol is dependent upon the pH value. Another supramolecular assembly approach employed the host–guest interaction between cyclodextrin (CD) and ferrocene, and formed a transparent supramolecular hydrogel by mixing poly(acrylic acid) (pAA) possessing β-CD as a host polymer with pAA possessing ferrocene as a guest polymer.[61] Redox stimuli could induce a sol–gel transition as well as control the self-healing property, as shown in Figure 5.8(G).

5.3.3 van der Waals (vdW) Interactions in Thermoplastic and Thermosetting Polymers

vdW interactions are weak intermolecular forces that play a significant role in the cohesion, adhesion, and mechanical properties of polymers. For instance, vdW forces enable a gecko to stick to walls and ceilings through the interactions between setae and the surface molecules.[62] Manipulating vdW forces in polymers has been explored in designing advanced materials for several decades.[63–65] However, only recently, vdW interactions were utilized to achieve multiple recoveries upon mechanical damage without external intervention in commodity copolymers.[66] The perturbation of vdW forces to an energetically unfavorable state upon mechanical damage causes interdigitated alternating/random copolymer motifs to self-recover to a more favorable energetic state. An example of this phenomenon was observed in poly(methyl methacrylate/*n*-butyl acrylate) p(MMA/*n*BA) copolymers and its derivatives within relatively narrow monomer molar ratios. These materials exhibit self-heal properties upon mechanical damage (Figure 5.9(A)) when the copolymer topologies are predominately alternating/random.

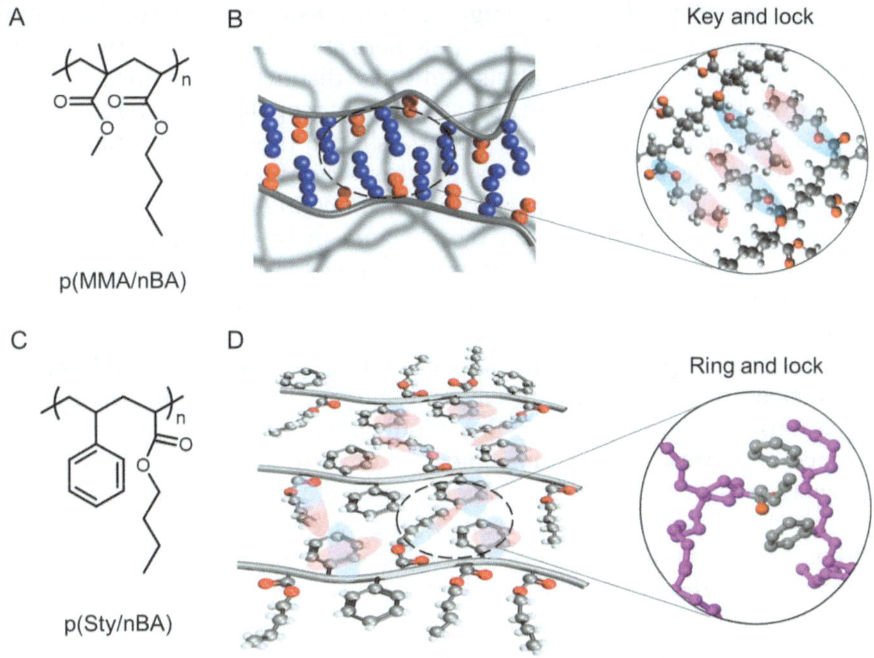

Figure 5.9 (A) Chemical structure of the poly(methyl methacrylate/*n*-butyl acrylate) (p(MMA/*n*BA) copolymer. (B) Key-and-lock associations in self-healable (p(MMA/*n*BA)) copolymers. (C) Chemical structure of poly(styrene/*n*-butyl acrylate) (p(Sty/*n*BA)) copolymer. (D) Ring-and-lock associations in self-healable p(Sty/*n*BA) copolymer. (A) and (B) Reproduced from ref. 66 with permission from American Association for the Advancement of Science, Copyright 2018. (C) and (D) Reproduced from ref. 67 with permission from American Chemical Society, Copyright 2023.

To understand the molecular events during the damage–repair cycle, a combination of molecular dynamics simulations and precise copolymerization was employed as a tool to determine the molecular interactions on self-healing. MD simulations revealed that the copolymer composition affects the formation of helix-like structures, leading to a viscoelastic response that favors self-recovery upon chain separation. This can be attributed to the 'key-and-lock' junctions among neighboring chains, stabilized by vdW interactions. Enhanced cohesive-energy density (CED) values further reflect the stabilization of neighboring copolymers attributed to vdW forces. Figure 5.9(B) illustrates how induced dipole interactions in alternating/random p(MMA/*n*BA) copolymers, driven by directional vdW forces, can enhance the CED equilibrium of entangled and side-by-side copolymer chains. The enhancement of vdW forces facilitates self-healing within the relatively narrow copolymer compositional range.

Using a similar self-healing strategy, commodity poly(styrene/*n*-butyl acrylate) p(Sty/*n*BA) copolymers also exhibit self-healing properties when employing alternating/random copolymer topologies (Figure 5.9(C)).[67]

The electronic effects of styrene's large electron density groups favor alternating topologies even under statistical copolymerization conditions. Two-dimensional correlation Raman analysis indicates distinct interactions between styrene rings in self-healable and non-self-healable copolymers, interchain face-centered (self-healable) and off-centered/T-shaped (non-self-healable) Sty–Sty interactions. Molecular dynamics simulations combined with spectroscopic and thermomechanical analysis attribute self-healing properties to favorable ring-and-lock associations driven by pi–sigma–pi (π–σ–π) vdW interactions between the π-orbitals of aromatic rings and the sigma components of aliphatic side groups (Figure 5.9(D)). Recent studies have shown that fluorophilic "sigma (σ)-lock" interactions may also facilitate self-healing.[68]

The healing of polymers can be accelerated by various external sources such as heat, light, moisture, or electric field.[69–71] For example, the healing of alternating/random p(MMA/*n*BA) copolymers can be accelerated in the presence of confined water molecules.[69] Competing vdW interactions that hinder H_2O from forming hydrogen bonds, forces *n*BA side groups to adopt L-shape conformations, generating stronger dipole–dipole interactions leading to shorter inter-chain distances compared to 'key-and-lock' associations in the absence of water. In addition to water, healing can also be accelerated by exposure to an electric field, which was demonstrated in self-healable poly(ionic liquid) copolymers.[70] When polymer chains are compressed upon damage, an alternating current electric field facilitates faster reorganization of energetically favorable interchain vdW interactions between dipolar hydrophobic groups, and accelerated recovery of vdW interactions and hydrogen bonding is responsible for faster self-healing.

The incorporation of exchangeable covalent bonds into self-healable motifs in commodity copolymers enables self-healing without intervention and reprocessing. One example is the covalently adaptable networks (CAN) of poly((2-acetoacetoxy)ethyl methacrylate/methyl methacrylate/*n*-butyl acrylate).[71] This copolymer network self-heals under ambient conditions and can be reprocessed by compression molding at 120 °C. Reprocessing is facilitated by the exchange reactions of dynamic cross-links, while self-healing involves the recovery of energetically favorable interchain vdW interactions, which parallels the reversible $E \rightleftharpoons Z$ isomerization of the C=C bonds of vinylogous urethane moieties. Similar studies were conducted in fluoro-containing copolymers.[72]

5.3.4 Shape Memory in Thermoplastic and Thermosetting Polymers

Creating mechanically robust materials with high strength and stiffness that can self-heal without requiring the application of high temperatures is a challenge. One approach to address this issue may involve using shape memory effects to facilitate self-healing while maintaining high strength and stiffness. In nature, the phenomenon of shape-memory-assisted self-healing is commonly observed in biological systems. For example, when wounds heal, the conformational entropic energy generated by the damage functions to close and seal the physical wound. This process subsequently enables essential chemical and physical processes for strength recovery without intervention.[73] In synthetic polymers, shape memory effects were first discovered and developed in the late 20th century.[74] Following this discovery, heat-shrinkable polyethylenes were developed that have found many industrial uses.[75,76]

Most shape memory polymers combine rubber elasticity and thermoplastic stiffness. Activating shape memory in polymers requires applying an external stimulus such as heat, light, or chemical redox.[77] This activation allows temporarily deformed polymers to revert to their permanent shape. Self-healing can be achieved when shape memory polymers are combined with dynamic and flexible macromolecular assemblies that increase mobility. Notably, changes in polymer chain conformation or contraction during self-healing are primarily driven by entropy. For example, thermoplastic polyurethane fibers composed of polycaprolactone diol (PCL-diol), 1,4-butanediol (BDO), hydroxyl modified spiropyran (SP), and hexamethylene diisocyanate (HDI) drawn from solution during polymerization can self-heal.[78] Figure 5.10(A) illustrates the chemical structure of this fiber in which the incorporation of soft and hard segments within a single polymer leads to microscale phase-separated morphologies (Figure 5.10(B)). The BDO in the hard-segment-rich domain cannot be disengaged under mechanical forces. Meanwhile, soft-segment-rich domains store entropic energy due to the unfolding of crystalline blocks and the elongation of low glass transition temperature (T_g) amorphous segments, resulting in decreased entropy. Upon heating, the chains in the soft-segment-rich domain become mobile enough to return to the randomized state, maximizing entropy. This released energy pulls crack surfaces together, leading to self-healing (Figure 5.10(C)).

Another approach to achieve autonomous self-healing while maintaining high strength and stiffness under ambient conditions is to leverage the continuous viscoelastic nature of the glass transition to gain sufficient molecular mobility. This can be achieved by measuring strain as a function of temperature in dynamic mechanical analysis (DMA) experiments (Figure 5.10(D)), which detects shape memory viscoelastic length transitions (VLTs).[79,80] Directional elongations occur in the polymer networks, displaying viscous behavior when approaching T_g. Subsequent retractions result from the released stored conformational entropic energy attributed to chemical or physical crosslinks and/or chain entanglements. Quantifying the correlation between VLTs and maximum stored strain, stress, and entropic energy density can predict relative shape memory capacities shown in Figure 5.10(E). The prediction plane illustrates the shape memory strength ranging from high-strain low-stress polymer networks to low-strain high-stress polymer networks. Higher VLTs, representing greater stored entropic energy, correspond to higher overall shape memory storage capacity, encompassing both stress and strain aspects. The correlation of VLTs in terms of maximum stored strain obtained in DMA measured maximum failure strains in static stress–strain measurements confirms a direct relationship between VLTs and shape memory capacity.

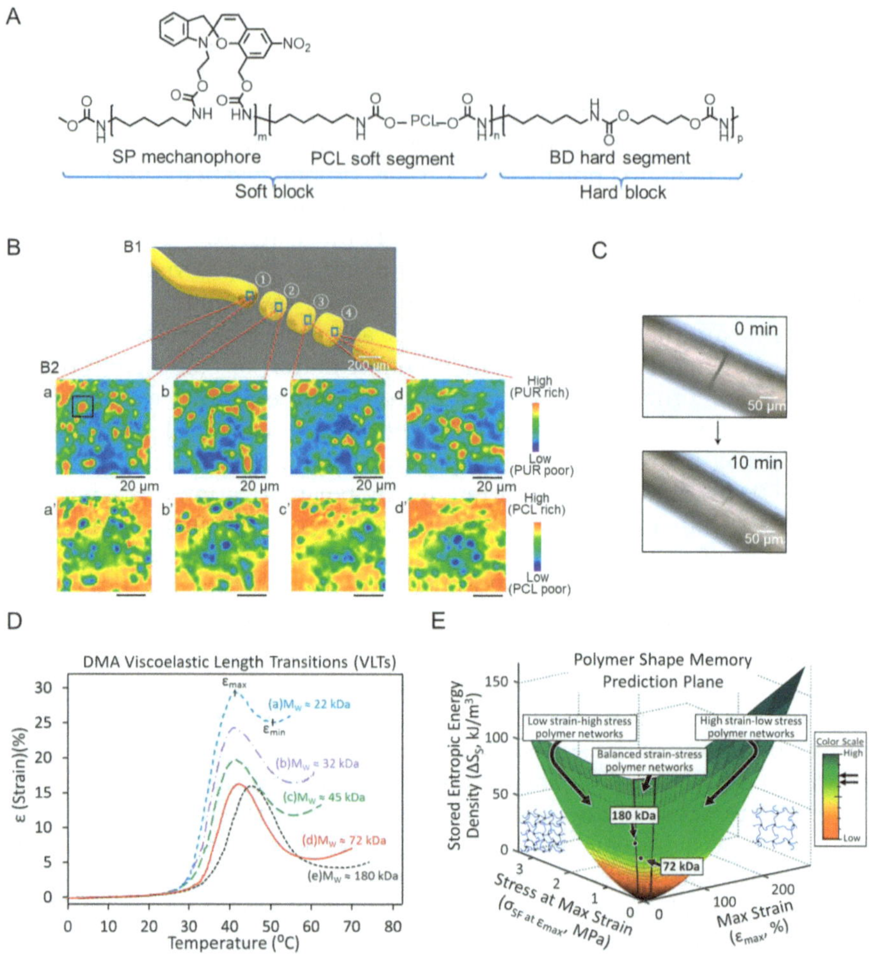

Figure 5.10 (A) Chemical structure of thermoplastic polycaprolactone–polyurethane fibers. (B1) Schematic representation of the synthesized thermoplastic polyurethane cross-section of a 200-mm thick fiber. (B2) IR images collected from the cross-sections of fiber. (C) Self-healing of fibers. (D) DMA strain curves showing the VLT measured from polymers with different molecular weights. (E) Shape memory polymer prediction plane. (A)–(C) Reproduced from ref. 78 with permission from Elsevier, Copyright 2018. (D) and (E) Reproduced from ref. 80, https://doi.org/10.1038/s41467-020-14911-y, under the terms of the CC BY 4.0 license http://creativecommons.org/licenses/by/4.0/.

5.4 Summary

In summary, although the field of multi-stimuli-responsive materials[81,82] in general, and self-healing in particular, has become more mature, there are many opportunities in this fascinating and continuously evolving field.[83,84]

The great challenge of imparting synthetic materials with the adaptability, responsiveness, and self-healing of synthetic 'living' polymers will be advanced through nano-micro-macro scale growth. Although the complexities of transient physico-chemical processes represent significant challenges, the future long-term payoffs will be substantial and will lead to a new generation of sustainable and environmentally compliant materials that will maintain their functions and appearance for extended times.

References

1. M. W. Urban, *Nat. Chem.*, 2012, **4**, 80.
2. D. Y. Wu, S. Meure and D. Soloman, *Prog. Polym. Sci.*, 2008, **33**, 479.
3. E. B. Murphy and F. Wudl, *Prog. Polym. Sci.*, 2010, **35**, 223.
4. R. P. Wool and K. M. O'Connor, *J. Appl. Phys.*, 1981, **52**, 5953.
5. P. G. de Gennes, *J. Chem. Phys.*, 1971, **55**, 572.
6. Y. H. Kim and R. P. Wool, *Macromolecules*, 1983, **16**, 1115.
7. K. M. Herbert, S. Schrettl, S. J. Rowan and C. Weder, *Macromolecules*, 2017, **50**, 8845.
8. Y. Yang and M. W. Urban, *Prog. Polym. Sci.*, 2015, **49-50**, 34.
9. P. J. Flory, *Proc. R. Soc. London, Ser. A*, 1956, **234**, 60.
10. F. Liu and M. W. Urban, *Prog. Polym. Sci.*, 2010, **35**, 3.
11. S. Wang and M. W. Urban, *Nat. Rev. Mater.*, 2020, **5**, 562.
12. K. Takeda, H. Unno and M. Zhang, *J. Appl. Polym Sci.*, 2004, **93**, 920.
13. A. S. Hay, *Adv. Polym. Sci.*, 1967, **4**, 496.
14. D. M. White and H. J. Klopfer, *J. Polym. Sci., Part A-1*, 1970, **8**, 1427.
15. H. Yang and A. S. Hay, *J. Polym. Sci.: Part A: Polym. Chem.*, 1993, **31**, 2015.
16. C. M. Chung, Y. S. Roh, S. Y. Cho and J. G. Kim, *Chem. Mater.*, 2004, **16**, 3982.
17. J. Paczkowski, *Polymeric Materials Encyclopedia*, ed. J. C. Salamone, CRC Press, Boca Raton, FL, 1996.
18. V. Ramamurthy and K. Venkatesan, *Chem. Rev.*, 1987, **87**, 433.
19. P. L. Egerton, E. M. Hyde, J. Trigg, A. Payne, P. Beynon, M. V. Mijovic and A. Reiser, *J. Am. Chem. Soc.*, 1981, **103**, 3859.
20. M. Hasegawa, T. Katsumata, Y. Ito, K. Saigo and Y. Litaka, *Macromolecules*, 1988, **21**, 3134.
21. M. Stevens and A. Jenkins, *J. Polym. Sci.*, 1979, **17**, 3675.
22. X. Chen, M. A. Dam, K. Ono, A. Mal, H. Shen, S. R. Nutt, K. Sheran and F. Wudl, *Science*, 2002, **295**, 1698.
23. Y. L. Liu, C. Y. Hsieh and Y. W. Chen, *Polymer*, 2006, **47**, 2581.
24. Y. L. Liu, C. Y. Hseih and Y. W. Chen, *Macromol. Chem. Phys.*, 2007, **208**, 224.
25. P.-G. de Gennes, *Hebd. Seances Acad. Sci., Ser. B*, 1980, **291**, 219.
26. P. P. Wang, S. Lee and J. P. Harmon, *J. Polym. Sci., Part B: Polym. Phys.*, 1994, **32**, 1217.
27. C. B. Lin, S. Lee and K. S. Liu, *Polym Eng. Sci.*, 1990, **30**, 1399.
28. C. C. Corten and M. W. Urban, *Adv. Mater.*, 2009, **21**, 5011.
29. M. Zako and N. J. Takano, *Intell. Mater. Syst. Struct.*, 1999, **10**, 836.

30. M. Motuku, U. K. Vaidya and G. M. Janowski, *Smart Mater. Struct.*, 1999, **8**, 623.
31. S. Kalista and T. Ward, *J. R. Soc. Interface*, 2007, **4**, 405.
32. S. J. Kalista, T. C. Ward and Z. Oyetunji, *Mech. Adv. Mater. Struct.*, 2007, **14**, 391.
33. D. Ramachandran, F. Liu and M. W. Urban, *RSC Adv.*, 2012, **2**, 135.
34. B. Ghosh and M. W. Urban, *Science*, 2009, **323**, 1458.
35. B. Ghosh and M. W. Urban, *J. Mater. Chem.*, 2011, **21**, 14473.
36. B. Ghosh and M. W. Urban, *J. Mater. Chem.*, 2012, **22**, 16104.
37. G. Deng, C. Tang, F. Li, H. Jiang and Y. Chen, *Macromolecules*, 2010, **43**, 1191.
38. Y. Amamoto, J. Kamada, H. Otsuka, A. Atahara and K. Matyjaszewski, *Angew. Chem., Int. Ed.*, 2011, **50**, 1660.
39. Y. Amamoto, H. Otsuka, A. Takahara and K. Matyjaszewski, *Adv. Mater.*, 2012, **24**, 3975.
40. J. Kamada, K. Koynow, C. Corten, A. Juhari, J. A. Yoon, M. W. Urban, A. C. Balazs and K. Matyjaszewski, *Macromolecules*, 2010, **43**, 4133.
41. J. A. Yoon, J. Kamada, K. Koynov, J. Mohin, R. Nicolay, Y. Zhang, A. C. Balazs, T. Kowalewski and K. Matyjaszewski, *Macromolecules*, 2012, **45**, 142.
42. J. Canadell, H. Groossens and B. Klumperman, *Macromolecules*, 2011, **44**, 2536.
43. K. Imato, M. Nishihara, T. Kanehara, Y. Amamoto, A. Takahara and H. Otsuka, *Angew. Chem., Int. Ed.*, 2012, **51**, 1138.
44. H.-G. Korth, *Angew. Chem.*, 2007, **119**, 5368.
45. R. C. Osthoff, A. M. Bueche and W. T. Grubb, *J. Am. Chem. Soc.*, 1954, **76**, 4659.
46. S. W. Kantor, W. T. Grubb and R. C. Osthoff, *J. Am. Chem. Soc.*, 1954, **76**, 5190.
47. P. Zheng and T. J. McCarthy, *J. Am. Chem. Soc.*, 2012, **134**, 2024.
48. R. P. Sijbesma, F. H. Bejer, L. Brunsveld, B. J. B. Fomer, J. H. K. Hirschberg, R. F. M. Lange, J. K. L. Lowe and E. W. Mejer, *Science*, 1997, **278**, 1601.
49. F. H. Beijer, H. Kooijman, A. L. Spek, R. P. Sijbesma and E. W. Meijer, *Angew. Chem., Int. Ed.*, 1998, **37**, 75.
50. S. H. M. Sontjens, R. P. Sijbesma, M. H. P. van Genderson and E. W. Meijer, *J. Am. Chem. Soc.*, 2000, **122**, 7487.
51. J. R. Fredericks and A. D. Hamilton, *Comprehensive Supramolecular Chemistry*, ed. J.-M. Lehn, Pergamon Press, New York, 1996, ch. 16.
52. P. Cordier, F. Tournilhac, C. Soulié-Ziakovic and L. Leibler, *Nature*, 2008, **451**, 977.
53. A. Phadke, C. Zhang, B. Arman, C.-C. Hsu, R. A. Mashelkar, A. K. Lele, M. J. Tauber, G. Arya and S. Varghese, *Proc. Natl. Acad. Sci. U. S. A.*, 2012, **109**, 4383.
54. Y. Chen, A. M. Kushner, G. A. Williams and Z. Guan, *Nat. Chem.*, 2012, **4**, 367.

55. U. S. Schubert, C. Eschbauner, O. Hien and P. R. Andres, *Tetrahedron Lett.*, 2001, **42**, 4705.
56. F. R. Kersey, D. M. Loveless and S. L. Craig, *J. R. Soc Interface*, 2007, **4**, 373.
57. M. Burnworth, L. Tang, J. R. Kumpfer, A. J. Duncan, F. L. Beyer, G. L. Fiore, S. J. Rowan and C. Weder, *Nature*, 2011, **472**, 334.
58. B. Lee, J. L. Dalsin and P. B. Messersmith, *Biomacromolecules*, 2002, **3**, 1038.
59. N. Holten-Anderson, M. J. Harrison, H. Brikedal, B. P. Lee, P. B. Messersmith, K. Y. Lee and J. H. Waite, *Proc. Natl. Acad. Sci. U. S. A.*, 2011, **108**, 2651.
60. L. He, D. E. Fullenkamp, J. G. Rivera and P. B. Messersmith, *Chem. Comm.*, 2011, **47**, 7497.
61. M. Nakahata, Y. Takashima, H. Yamaguchi and A. Harada, *Nature Commun.*, 2011, **2**, 511.
62. K. Autumn, M. Sitti, Y. A. Liang, A. M. Peattie, W. R. Hansen, S. Sponberg, T. W. Kenny, R. Fearing, J. N. Israelachvili and R. J. Full, *Proc. Natl. Acad. Sci. U. S. A.*, 2002, **99**, 12252.
63. C. J. Drummond and D. Y. C. Chan, *Langmuir*, 1997, **13**, 3890.
64. W. Wang, *Nat. Rev. Phys.*, 2002, **4**, 504.
65. Z. Huang, A. Alharb, W. Mayer, E. Cuniberto, T. Taniguchi, K. Watanabe, J. Shabani and D. Shahrjerdi, *Nat. Commun.*, 2020, **11**, 3029.
66. M. W. Urban, D. Davydovich, Y. Yang, T. Demir, Y. Zhang and L. Casabianca, *Science*, 2018, **362**, 220.
67. S. Gaikwad and M. W. Urban, *J. Am. Chem. Soc.*, 2023, **145**, 9693.
68. S. Gaikwad and M. W. Urban, *Angew. Chem., Int. Ed.*, 2024, **63**, e202405504.
69. D. Davydovich and M. W. Urban, *Nat. Commun.*, 2020, **11**, 5743.
70. Q. Liu, S. Wang, Z. Zhao, J. Tong and M. W. Urban, *Small*, 2022, **18**, 2201952.
71. S. Wang, L. Li, Q. Liu and M. W. Urban, *Macromolecules*, 2022, **55**, 4709.
72. S. Wang, L. Li and M. W. Urban, *ACS Appl. Polym. Mater.*, 2022, **4**, 9360.
73. O. Speck, M. Schlechtendahl, F. Borm, T. Kampowski and T. Speck, *Beilstein J. Nanotechnol.*, 2018, **9**, 175.
74. L. B. Vernon and H. M. Vernon, *US Pat.*, 2234993, 1941.
75. W. C. Rainer, E. M. Redding, J. J. Hitov, A. W. Sloan and W. D. Stewart, *US Pat.*, 3144398, 1964.
76. R. J. Perrone, *US Pat.*, 3326869, 1967.
77. A. Lendlein and O. E. C. Gould, *Nat. Rev. Mater.*, 2019, **4**, 116.
78. Y. Yang, D. Davydovich, C. C. Hornat, X. Liu and M. W. Urban, *Chem*, 2018, **4**, 1928.
79. C. C. Hornat, Y. Yang and M. W. Urban, *Adv. Mater.*, 2017, **29**, 1603334.
80. C. C. Hornat and M. W. Urban, *Nat. Commun.*, 2020, **11**, 1028.
81. M. W. Urban, *Stimuli-Responsive Materials; From Molecules to Nature Mimicking Materials Design*, Royal Society of Chemistry, Cambridge, UK, 2016.
82. S. Wang and M. W. Urban, *Chem*, 2023, **9**, 1362.
83. J. Liu and M. W. Urban, *Langmuir*, 2024, **40**, 7268–7285.
84. J. Liu and M. W. Urban, *Mater. Horiz.*, 2025, DOI: 10.1039/d5mh00710k.

Subject Index